ARRL's

HF DIGITAL
HANDBOOK

D0861735

Steve Ford, WB8IMY

Published by
ARRL—the national association for Amateur Radio
Newington, CT 06111 USA

CONTENTS

FOREWORD

As you read the first chapter of *ARRL's HF Digital Handbook*, you'll discover that the evolution of amateur HF digital communication has been nothing short of phenomenal—especially in recent years. Hams exploited RTTY after World War II and relied on it for decades to move text data throughout the world. Little changed until amateurs finally got their hands on affordable computer technology in the early '80s. HF digital operators quickly embraced AMTOR and the race for HF digital innovation began. It continues to this day.

ARRL's HF Digital Handbook is not intended to be a sentimental retrospective, however. Instead, the book serves as a guide to understanding the most active HF digital modes in use today. These include RTTY (its popularity continues after more than 50 years), PSK31, PACTOR, Clover, G-TOR and Hellschreiber.

There is something within *ARRL's HF Digital Handbook* for every reader. Beginners will enjoy the practical advice and relaxed writing style. More advanced operators will appreciate the concise explanations of coding techniques and modulation methods, as well as the ample resource information.

With rapid advances in information technology, HF digital communication has become a moving target. New techniques are being developed on a regular basis as more powerful computers find their way into ham stations. *ARRL's HF Digital Handbook* is, therefore, a snapshot look at the current state of the art. The challenge to you will be to keep up to date with new modes and methods as they rapidly unfold!

David Sumner, K1ZZ

Executive Vice President
January 2001

Chapter ONE

Welcome to the World of HF Digital

Amateur Radio has entered the 21st Century. It is remarkable when you realize that in less than 100 years amateur communication has evolved from crude spark-gap technology to digital signal processing. Where hams once had to choose between voice and CW, we now enjoy a broad range of communication choices from television (slow and fast scan) to spread spectrum. Where we were once limited to frequencies below 1500 kHz, we now communicate on bands from medium waves to microwaves.

Amateur digital communication has also evolved. Take a look at the time line in **Figure 1-1**. From the end of World War II until the early '80s, radio telegraphy, better known as *RTTY*, was the only HF digital mode available to amateurs. Then, in 1983, AMTOR made its debut, coinciding with the rising popularity of personal computers. It was the first amateur digital communication mode to offer error-free text transmission.

From the early '80s the rate of change quickened to a fast gallop. Packet radio emerged by the middle '80s and, for a time, reigned as the

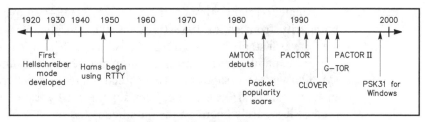

Figure 1-1—The Amateur Radio digital time line.

most popular form of amateur digital communication. As microprocessor technology became more sophisticated, we saw the rise of modes such as Clover, PACTOR and G-TOR that were capable of error-free exchanges under marginal band conditions (weak signals, interference and so on). In the late '90s there was an invention that harnessed personal computer technology to create a new digital mode for casual operating: PSK31. In the first years of the 21st Century the evolution has continued with more HF digital modes such as MFSK16.

Another powerful agent for change in the amateur digital community has been the Internet. With the possible exception of the personal computer, nothing has had such a profound impact on the hobby. The Internet allowed amateurs to exchange information, including software, quickly and reliably. The development of the World Wide Web created a new kind of publishing arena where anyone could launch a personal Web site to promote their favorite aspects of Amateur Radio. As new ideas and applications flowed across the network, computers became essential components of modern ham stations. (By 2000, for example, more than 90% of all ARRL members used computers in their amateur stations, and more than 80% had Internet access.) The explosive growth of the Internet has had a substantial impact on amateur digital communication, HF and otherwise.

Let's take a brief look at current amateur HF digital communication modes and where they stand today.

RTTY

RTTY is the granddaddy of HF digital, although its popularity has been seriously undercut by PSK31 (more about that mode in a moment). RTTY is primarily used for casual conversation, and it's still the mode of choice for digital contesting and DXing.

AMTOR

Amateur Teleprinting Over Radio—AMTOR—enjoyed widespread popularity from about 1983 through 1991. Its distinctive *chirp-chirp* sound was a staple on the HF bands. Hams made ample use of its error-free text capability, even setting up automatic AMTOR mailbox operations (MBOs) where messages could be stored for later retrieval from anywhere in the world. AMTOR has since been superceded by faster, more versatile modes. It is rarely heard on the ham bands today.

PACKET

Although packet technology had been in existence since the early '70s, hams embraced it with gusto in the middle '80s. (Personal computers, again, were the driving force.) Packet is an error-detecting mode, which means that it is capable of communicating error-free information, including binary data (for images, software applications, etc). The problem with packet, as far as HF communication is concerned, is that it requires strong, "quiet" signals at both ends of the path to function efficiently. Packet doesn't tolerate signal fading, noise or interference, which makes it a poor choice for the chaotic world of HF.

Packet was much more successful on VHF and UHF. For nearly a decade packet flourished, primarily on the 2-meter band. Hams reveled in their newfound ability to send e-mail messages throughout the nation and the world, but the advent of affordable Internet access soon brought the party to a grinding halt. As the Internet skyrocketed to prominence, traditional packet declined just as rapidly. Bulletin board stations closed, networks collapsed and sales of packet equipment slowed to a trickle.

VHF packet retains a foothold today thanks to DX spotting networks (known as DX PacketClusters) and the moderate popularity of the Automatic Position Reporting System (APRS), a marriage of amateur packet radio and global positioning satellite technology. Despite its poor performance, HF packet remains stubbornly alive, though most of the activity is concentrated in overseas operations in Third World nations where the affordability of packet equipment is still a strong drawing card.

PACTOR

PACTOR strolled onto the telecommunications stage in 1991. It combined the best aspects of packet (the ability to pass binary data, for example) and the robust error-free nature of AMTOR. It was eagerly embraced by HF digital equipment manufacturers and became the number-one HF digital communication mode in a remarkably short period of time. PACTOR was popular for mailbox operations and other forms of message handling. As with packet, the Internet caused a serious decline in PACTOR activity, but it remains the most popular of the error-free modes.

PACTOR II debuted in the mid '90s as a rival to Clover, and the two have been doing battle for the hearts, minds and pocketbooks of HF communicators (commercial and amateur) ever since. Like Clover, PACTOR II uses DSP techniques and complex data coding to achieve extraordinary performance. Also like Clover, the necessary equipment is quite expensive, which has slowed PACTOR II's acceptance in the ham community. In 1999 the creators of PACTOR II unveiled a pared-down processor that offered the same performance, but at a more attainable cost.

CLOVER

Clover was unveiled in 1993 by the HAL Communications Corporation. It was one of the first HF digital modes to use sophisticated data coding, coupled with complex modulation schemes and digital processing technology, in an effort to overcome the vagaries of HF. Clover promised, and delivered, impressive performance even in the face of weak signals and terrible band conditions. This performance initially came at a stiff price—one that few hams could afford. As you'd expect, the high cost of Clover technology dampened enthusiasm in the beginning. Price reductions later in the decade, and the introduction of Clover II, helped the mode retain a dedicated following.

G-TOR

G-TOR was the brainchild of Kantronics, a digital communication equipment manufacturer. It was yet another high-performance mode, although not as costly as Clover or PACTOR II. Like both of the former, however, G-TOR was *proprietary*. That means that it is only available in equipment manufactured by Kantronics. Coming several years after the appearance of PACTOR, G-TOR never really captured the attention of HF digital operators. It is somewhat uncommon on the ham bands today as a result.

PSK31

PSK31 could be viewed as a high-octane cousin of RTTY. It is not an error-free digital mode, but it offers excellent weak-signal performance. Peter Martinez, G3PLX, the same person who created AMTOR, invented PSK31. For a few years PSK31 languished in obscurity because special DSP hardware was necessary to use it. But in 1999 Peter designed a version of PSK31 that needed nothing more than a common computer soundcard. It was a simple piece of software that ran under *Windows* and used the soundcard as its interface to the transceiver. Peter made the software available at no cost on the Internet, and that was like tossing a lighted match into a can of gasoline. Within a few months PSK31 exploded in the HF digital community. As this book was being written, PSK31 was emerging as a possible successor to RTTY for casual ragchewing, contesting and DXing.

HELLSCHREIBER

Hellschreiber is not a new mode (it was pioneered in the 1920s and '30s by Rudolf Hell), but a number of hams are beginning to discover its possibilities. Unlike all of the other modes we've discussed so far, Hellschreiber is *visual*. That is to say, the signals "paint" the text on your screen much in the same sense that a television or fax signal paints an image.

One variation of Hellschreiber known as *Feld-Hell* works its magic by keying a CW transmitter ON for every black portion in a text character, and OFF for every white space. Timing is critical. See **Figure 1-2** for an example of Feld-Hell signal reception. Feld-Hell has drawn some interest among low power (QRP) operators because you can operate with simple (but stable) CW transmitters. Most Feld-Hell operation, however, is done using SSB transceivers using on/off tone "keying" to accomplish the same result. Feld-Hell is the most popular of the Hellschreiber modes.

You can also send text imagery by using different *frequencies* (tones) to represent the black and white areas. This version of Hellschreiber is called *Multi-Tone Hell*, or simply MT-Hell. There are several variations of MT-Hell in use (see Chapter 8).

MT-63

MT-63 was invented by Pawel Jalocha, SP9VRC. It is a keyboard-to-keyboard "live" mode operationally similar to RTTY and PSK31. With MT-63, however, the data components are spread over 64 different tones! This allows a tremendous amount of redundancy, assuring good reception even when as much as 25% of the data has been obliterated by noise, fading or interference. Thanks to its modulation structure, MT-63 offers excellent performance under poor conditions, even rivaling Clover and PACTOR II.

There is a certain amount of controversy surrounding MT-63 in the

Figure 1-2—An example of Feld-Hell in action.

Popular HF Digital Frequencies

Band (Meters)	Frequencies (MHz)
10	28.070—28.130
12	24.920—24.930
15	21.060—21.099
17	18.100—18.110
20	14.060—14.099[1]
30	10.120—10.150
40	7.060—7.099[2]
80	3.580—3.640

Notes

[1]This is the most active HF digital band

[2]Digital operators should avoid interfering with operators in ITU Regions 1 and 3, as well as Alaska, Hawaii and Puerto Rico, who may have phone privileges in this portion of 40 meters.

amateur community. Its signal is quite wide (1 kHz). With crowded conditions in the HF digital subbands today, the movement has been toward narrow signals. PSK31, for example, is only about 31 Hz wide. MT-63 seems to run counter to this trend. Finally, there are legal issues involving the complex MT-63 modulation scheme. As this book went to press, the Federal Communications Commission had not declared MT-63 a legal mode for US-licensed amateurs.

MFSK16

This mode was brought to the forefront through an article by Murray Greenman, ZL1BPU, that appeared in the January 2001 issue of *QST*. MFSK16 uses several different tones to communicate in difficult conditions, yet still remaining within a bandwidth of about 300 Hz. The mode was developed as a joint effort between ZL1BPU and Nino Porcino, IZ8BLY. It was Nino's *Stream* software that introduced MFSK16 to the ham community at large.

ENOUGH TALK!

You have a lot of digital modes to explore—a virtual smorgasbord of choices! The purpose of this book is to give you resources to get started in the most active modes: RTTY, PSK31, PACTOR, Clover, G-TOR and Hellschreiber. The first place to begin is by assembling your HF digital station. Turn the page and let's get started!

Chapter TWO

Assembling Your HF Digital Station

In the old days of HF digital, back when RTTY was your only option, setting up a station wasn't a trivial exercise. You had to make room for a bulky mechanical teletype machine, cobble together an interface to your radio to actuate the teletype, and install an oscilloscope to help you tune the signal. You sent text by typing on the teletype's awkward "green keys," and read the other station's replies on sheets of yellow paper. Some miss the good old days of clattering teletypes and dripping oil, but I'm not counted among them!

Thanks to personal computers and microprocessor technology it is infinitely easier to assemble an HF digital station these days. You have several choices available to you, depending on the resources you already own and how much you want to spend.

DIGITAL TO ANALOG—AND BACK AGAIN

Data exists in your computer in the form of changing voltages. Five volts might represent a binary "1" while zero volts may represent a binary "0." But a radio can't transmit changing voltages—at least not without a little help.

If you stroll up to me and begin conversing in French, all you'll get is a blank look (or perhaps a bewildered grimace). But if my wife begins whispering English translations in my ear, you may see a glimmer of recognition. (She knew her college French would come in handy someday.)

In the case of digital communication, our translator is a modulator/demodulator, otherwise known as a *modem*. A modem takes the data from your computer and translates it into shifting audio tones. One tone represents a binary "1" and another represents a binary "0." (Digital operators refer to these as *mark* and *space* tones.) The difference in their frequencies is their *shift*. The HF RTTY standard is a 170-Hz shift, although this has "migrated" to a 200-Hz shift thanks to accommodations made for packet and PACTOR modems. Regardless of the frequency, the rest is easy. Feed the tones to an SSB transceiver and you're in business!

This basic setup is called *AFSK*, or audio frequency-shift keying. Your transceiver manual may refer to it as *FSK*, and this can be a little confusing. True FSK involves changing the frequency of your rig's

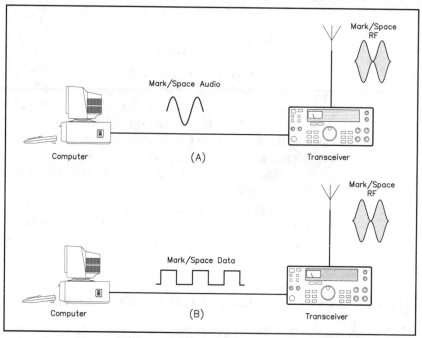

Figure 2-1—When you use Audio Frequency Shift Keying (AFSK), your computer sound card or multimode processor generate audio tones, which you feed to your transceiver to create RF (A). With Frequency Shift Keying (FSK), however, your computer or processor only generates data pulses (B). When fed to an FSK-capable transceiver, these data pulses signal the radio to transmit the Mark or Space signals. Whether you're operating FSK or AFSK, the results are essentially the same. The advantage of FSK is that many radios only implement narrow IF receive filtering when in the FSK mode. This is a big plus for DXers or contesters. On the other hand, FSK is strictly a binary ON/OFF system. HF digital modes such as PSK31, CLOVER, MFSK16 and Hellschreiber cannot run under conventional FSK. You must operate these modes in AFSK.

master oscillator in sync with the data from your computer. Few operators use FSK these days, although there may be some advantages in doing so. Most hams prefer AFSK. With either method, the result on the receiving end is the same (see **Figure 2-1**).

And what about reception? The modem waiting patiently at the other end of the path is equipped with tone decoders and very sharp audio filters. It will respond only to the proper tones *if* they are shifted by the correct amount. Off-frequency tones are ignored, and tones separated by incorrect shifts never make it past the filters. But when the tones hit the targets, the modem instantly converts them into data pulses—which soon wind up in your computer.

What I've just described is a simple, two-tone system. If you're talking about the more advanced digital modes, things can get more complicated. More than one tone may be used, or there may be some legerdemain with phasing or frequency. The essential principle remains the same: Digital to analog (audio), then back again.

Basic Modems

You can do RTTY and AMTOR on the cheap with modems consisting of a couple or just a handful of parts (**Figure 2-2**). With this type of simple modem you can use shareware programs such as *HamComm* (see the "Resources" section of this book) and get on the air in no time. You can build these modems yourself, or buy them ready to go. There are simple stand-alone modems such as the MFJ-1214 and others that are built directly into DB-25 shells that plug into your computer's COM or LPT (printer port). The TigerTronics BP-2M is a good example of a small multimode modem that offers impressive performance on RTTY or AMTOR for less than $100. MFJ Enterprises makes a device known as the MFJ-1213 that will put you on RTTY for less than $50.

Soundcards as HF Modems

You may already be the proud owner of an HF modem—your PC sound card! Sound cards do exactly what modems do: Convert data to audio and audio to data. With the proper software running in the background, your PC can become a high-performance digital communication machine.

In recent years PCs have become increasingly powerful and soundcards are as ubiquitous as hard drives. As a result, there appears to be a growing movement in the ham community toward using soundcards as radio

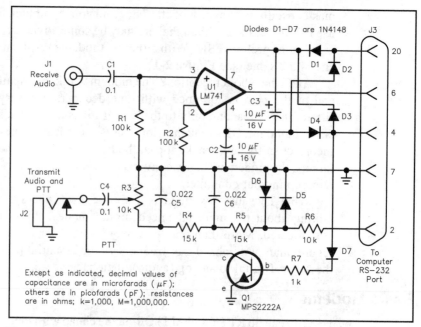

Figure 2-2—The classic HamComm modem interface. You can build the interface in less than an hour and get on the air with RTTY and a number of other modes. The *HamComm* software is available free on the Web. See the "Resources" section of this book. RadioShack part numbers are shown below in parentheses. Note that R3 is used to adjust the transmit audio level.

C1, C4—0.1 μF ceramic disk capacitors (272-135)
C2, C3—10 μF, 16 V electrolytic capacitors (272-1013)
C5, C6—0.022 μF ceramic capacitors (272-1066)
D1-D7—1N4148 diodes (276-1122)
J1—RCA phono jack
J2—Miniature ⅛-inch stereo phone jack
J3—25-pin DB-25 connector
Q1—MPS2222A transistor (276-2009)
R1, R2—100 kΩ resistors, ¼ W (271-1347)
R3—10 kΩ potentiometer (271-282)
R4, R5—15 kΩ resistors, ¼ W (271-1337)
R6—10 kΩ resistor, ¼ W (271-1335)
R7—1 kΩ resistor, ¼ W (271-1321)
U1—LM741 Op Amp (276-007)

Many modern laptops no long offer COM ports, but they do have USB ports. This neat little device from D-Link known as the DSB-S25 is a USB-to-serial-port converter. Just plug one end into your laptop USB port and the other end (with the male DB-25 connector) functions as a serial port for keying your transceiver or connecting to a multimode processor. The device costs about $60. More information is available on the Web at *www.dlink.com*.

Among the easiest and most affordable ways to get started with RTTY are the Tigertronics BP-2M or MFJ-1214 modems. These devices plug into any available COM port. Just run the software provided and you're in business.

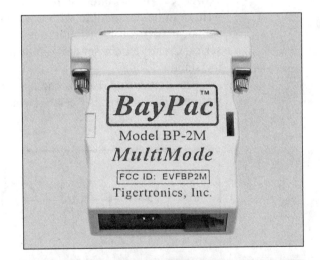

"modems." Will this spell the end of outboard interfaces and multimode processors? Time will tell!

Brian Beezley, K6STI, was among the first to use the soundcard as a high-performance modem with his *RITTY* software. Others quickly followed. When PSK31 exploded onto the scene in the late '90s, it was implemented *entirely* in software using soundcards. The same is now true for the *Hellschreiber* modes. Believe it or not, you can now operate RTTY, AMTOR, PACTOR, PSK31 and Hellschreiber using your soundcard as your modem (signal connections are shown in **Figure 2-3**). The only external hardware you need is a simple switching circuit to key your radio (see **Figure 2-4**). Alternatively, you can build a more elaborate interface that allows you to fine-tune the transmit and receive audio levels (**Figure 2-5**).

Of course, there is a catch or two. The first is that your computer must be capable of running the processor-intensive software. For PCs we're talking about, at minimum, a Pentium-class machines such as a Pentium 133. If you have a love affair with Macs, you'd better invest in a PowerPC. The second catch is the soundcard. HF digital software is written to be compatible with a wide variety of soundcards, but not *all* soundcards. If you're in doubt stick with the Creative Labs "SoundBlaster" series. Creative Labs set the soundcard standard years ago and it remains the standard today.

I'm often asked if hams can use laptop computers in their HF digital stations. The answer is "yes," if your laptop has sound capability with

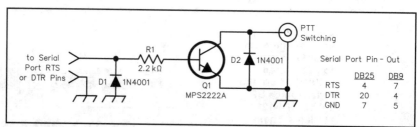

Figure 2-3—This simple circuit will allow your computer's COM port to key your transceiver when you're using soundcard software. You can choose to connect to either the RTS or DTR pins at the COM port. Some software packages key the transceiver by sending a logic "high" to one pin or the other. PSK31 software sends the logic high to both pins. RadioShack part numbers are shown in parentheses.

D1—1N4001 diode (276-1101)
R1—2.2 kΩ resistor, $\frac{1}{4}$ W (271-1325)
Q1—MPS-2222A transistor (276-2009)

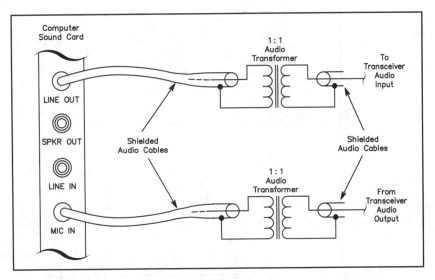

Figure 2-4—Sound card audio interfacing is often straightforward. In this example we have a full-featured sound card with Line In and Line Out ports, as well as a microphone input and speaker output. The Line Out jack is ideally suited for feeding audio to your transceiver, but you can use the speaker output if you are careful to adjust the output to a low level. Many HF digital operators prefer to use the Line In for audio *from* the transceiver, but this port isn't always available (especially on laptops). If so, use the microphone input. Note that 1:1 audio isolation transformers are used in both audio lines to reduce hum and noise.

external speaker and microphone jacks. You'll need to use an interface to reduce the audio level between the radio and the microphone jack. You can build the interface shown in Figure 2-5, or purchase pre-assembled interfaces such as the RIGblaster by West Mountain Radio. Concerning the soundcard, most laptop manufacturers adhere to the Creative Labs SoundBlaster convention, so you shouldn't have any problems in that department. There are no guarantees, though!

Multimode Processors

Until soundcard software appeared on the market, the most common HF digital interface was the *multimode processor*, or what some refer to as a *multimode TNC*. This device offers several HF digital communication modes in a single unit that usually sits alongside your radio. The exception is the HAL Communications P38 that comes in the form of a

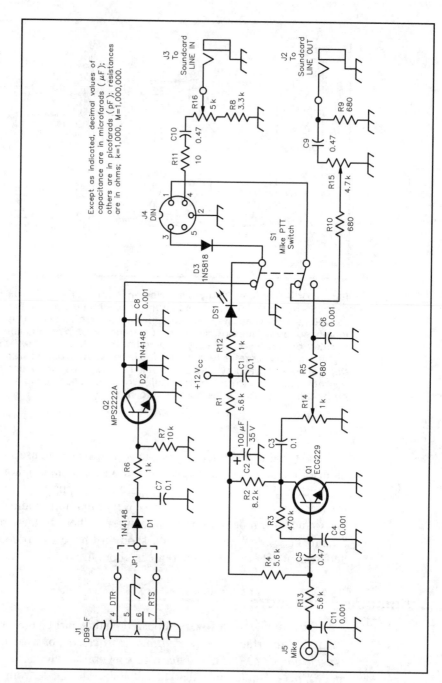

Figure 2-5—(caption on next page).

Figure 2-5—A versatile soundcard interface designed by Salvador Esteban, EB3NC. It provides transmit/receive switching, allows you to set both transmit and receive audio levels, and even includes a microphone jack with a preamp!

C1, C7—0.1 µF ceramic
C2—100 µF 25V electrolytic
C3—0.1µF polyester
C4, C6, C8, C11—0.01µF ceramic
C5, C9, C10—0.47 µF polyester
D1, D2—1N4148 diode
D3—1N5818 diode
DS1—LED (red)
R1, R4, R13—5.6 kΩ ¼ W
R2—8.2 kΩ ¼ W
R3—470 kΩ ¼ W

R5, R9, R10—680 Ω
R11—10 Ω
R6, R12—1 kΩ ¼ W
R7—10 kΩ ¼ W
R8—3.3 kΩ ¼ W
R14—1 kΩ trimmer
R15—4.7 kΩ trimmer
R16—5 kΩ linear taper
Q1—ECG229 transistor
Q2—MPS-2222A transistor

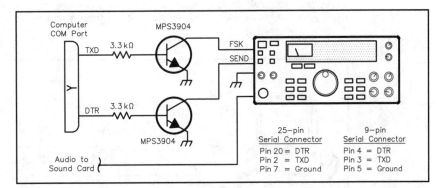

Figure 2-6—Is it possible to run FSK with a sound card setup? You bet! Your RTTY software can generate the Mark/Space data at the computer's serial (COM) and this dual-transistor circuit will convert the serial data into Mark/Space pulses for your radio. The audio from your radio is fed to the Line or Microphone inputs of your sound card. Before you try this, make sure your RTTY sound card software supports FSK.

circuit "card" that fits *inside* your PC.

Multimode processors offer several HF digital modes, plus a cornucopia of features such as message mailboxes, remote control and much more. The processors are relatively easy to hook up between your computer and your transceiver (see **Figure 2-7**). The computer uses terminal software to "talk" to the processor. The processor manufacturer usually provides this software, but there are third-party sources as well. Between the processor and the radio you need to run cables for receive audio, transmit/receive switching and transmit audio.

The elegant solution to sound-card interfacing and keying is available in the form of the RIGblaster from West Mountain Radio, 18 Sheehan Ave, Norwalk, CT 06854; tel 203 853 8080; *K1UHF@westmountainradio.com*; *www.westmountainradio.com.* This neat little box takes care of transceiver keying, microphone switching, audio level matching and so on.

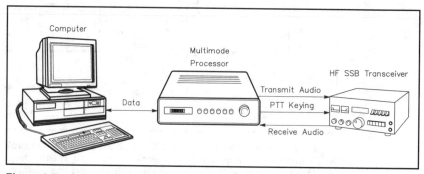

Figure 2-7—A multimode communication processor connects to the COM port of your computer and to your transceiver. Use short, shielded cables for all connections.

But with so many soundcard-based modes available, what are the advantages of purchasing an outboard processor?

• You can use almost *any* computer with a multimode processor. The processor merely asks the computer to send and receive information. This is a easy task for even the oldest machines. As long as the computer has terminal software that can communicate with the processor—or if the computer can run the terminal software provided by the manufacturer—you're in business. You could use an "ancient" 286 or 386 PC, a used $200 laptop or whatever.

• Many multimode processors offer features that are not presently

If you want to try Clover as well as RTTY and PACTOR, the solutions are HAL Communications P38 or P38DX. The P38 plugs into your PC buss slot while the P38DX is an outboard unit. Both products sell at less than $400.

available with soundcard-based software. Mailboxes and automatic, stand-alone operations come to mind. Your processor can be set up to receive and process messages without the need for supervision by your computer. You can ignore the processor completely and use your PC to do other work.

• Clover, G-TOR and PACTOR II are available *only* with multimode processors. If you wish to try these modes, you must purchase a HAL Communications unit (for Clover), a Kantronics KAM (for G-TOR) or an SCS PTC-IIe (for PACTOR II). All of these products will also operate RTTY, PACTOR, AMTOR and, in the case of the SCS PTC-IIe, PSK31.

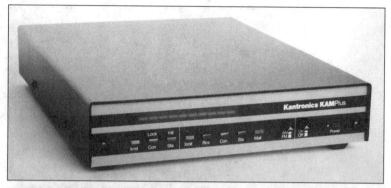

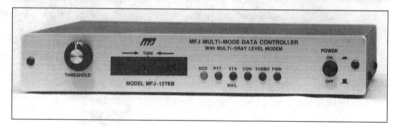

Popular multimode communication processors include the Kantronics KAM Plus, the Timewave PK-232/DSP and the MFJ-1278B.

THE TRANSCEIVER

Just about any modern SSB transceiver will do double duty as your HF digital radio. However, if you plan to dabble in the handshaking modes such as AMTOR, PACTOR, PACTOR II, Clover or G-TOR you want a radio that can jump from transmit to receive (and back again) very quickly—usually in less than 30 milliseconds. When a link is established and your computer sends a chunk of data, you must be ready to receive

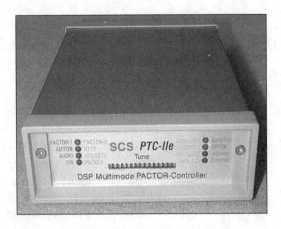

The SCS PTC-IIe is the only multimode communication processor that offers PACTOR II. The original PTC-II was priced at nearly $1000, which made it slow to catch on in the ham community. The PTC-IIe, however, sells for less than $700.

data from the receiving station right away. In other words, your radio has to exit the transmit mode and be ready to receive by the time the data signal comes screaming back at the speed of light. Fortunately, most radios manufactured after 1985 can perform this trick reasonably well. I even managed to get an old 1968-vintage Drake TR-4 transceiver to run AMTOR, although it sounded like it was about to shake itself apart! When in doubt, however, check the transceiver specifications carefully.

For RTTY, PSK31, MFSK16 or Hellschreiber any SSB radio, modern or otherwise, is sufficient as long as it is reasonably stable. If you intend to operate PSK31, however, you'll need a rig that can tune in very small steps—somewhere in the vicinity of 1 Hz. Quite a few late-model radios can do this, but check the specs before you buy. If the transceiver you already own can only tune in, say, 10-Hz steps, you're still in the PSK31 ballgame. You'll just have to rely on the radio's receiver incremental tuning (RIT) control and the automatic frequency compensation in the PSK31 software to get you on target.

And when it comes to RTTY, PSK31, MFSK16 or Hellschreiber you're not engaging in the handshake dance. All transmit/receive switching is controlled manually. This means that you don't have to be too concerned about how fast your rig jumps from transmit to receive.

The only remaining concern with RTTY or PSK31 is output power. When you transmit with these modes you're asking your rig to produce full output for several minutes at a time—maybe longer if you're a slow typist. This is known as a *100% duty cycle*. Most transceivers won't tolerate this kind of punishment; they're designed for the lower duty cycles of SSB or CW. So, when you're running RTTY or PSK31 it pays to reduce your output. Some manufacturers recommend a reduction of about 50%.

Accessory Jacks

The nice thing about modern transceivers is that most are equipped with handy rear panel accessory (or auxiliary) jacks. These multipin jacks usually feature a transmit/receive keying line, a fixed-level receive-audio output and a transmit audio input. These three connections make it astonishingly easy to set up your HF digital station.

The fixed-level receive-audio output is exactly what your soundcard, modem or processor needs to function efficiently. Because it is a fixed, continuous audio source, you can turn down the audio gain on your transceiver without affecting the performance of the data interface. Personally, I enjoy hearing the sound of the data signal while I am carrying on a conversation, but to each his or her own!

A keying line at the accessory jack means that your modem or processor can automatically switch your radio from transit to receive as necessary. You won't have to do the switching manually.

And a transmit audio input is tailor-made for HF digital signals. Unlike your microphone jack, it expects the higher audio levels generated by your soundcard or modem.

If you are using a transceiver that does not include an accessory jack, all is not lost. You can take the receive audio from your external speaker jack (**Figure 2-8**), although you'll need to take care not to overdrive the soundcard, modem or processor. The transit/receive keying line can be wired in through your microphone connection, as can your transmit audio. You'll need an attenuator to reduce the audio level, though (**Figure 2-9**).

Filtering

Most communications processors and soundcard programs implement some form of audio filtering to reduce the signal chaos and make it easier to decode the data you want. Even so, such filtering is occasionally insufficient, particularly when the bands are very crowded (during contests, for example).

If your transceiver has a built-in audio filter—many modern rigs feature digital signal processing (DSP) filters—you can use it to dramatically reduce interference and improve your data throughput. You can also purchase an external audio filter to achieve the same result.

Filtering in the intermediate frequency (IF) stages of your radio is probably most effective at keeping adjacent frequency interference at bay. HF digital operators typically use 500-Hz IF filters for all modes except packet and MT-63. If you are shopping for a new radio, look for a model that offers 500-Hz filters, either as options or standard equipment.

Check the IF filter descriptions carefully, however, and read the fine

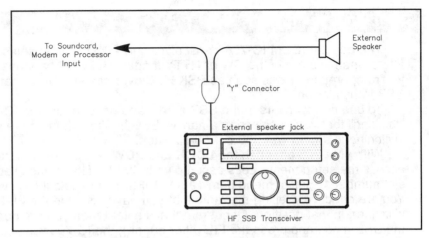

Figure 2-8—With a Y connector (RadioShack 274-304) and a small external speaker (RadioShack 40-1991) you can split the audio from your external speaker jack and send it to a modem, soundcard or multimode communication processor. The only problem is that you must set your audio gain control to a specific level and leave it there. One possible solution may be to install a small audio amplifier (RadioShack 277-1008) between the Y connector and the external speaker. You could still leave the transceiver's audio gain at a fixed position while adjusting the audio amplifier for a comfortable listening level at the external speaker.

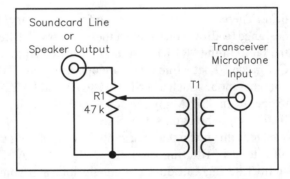

Figure 2-9—If you are using a soundcard as your HF digital modem and your transceiver does not have an accessory jack, you may have to feed your transmit audio directly into the microphone jack. Use the attenuator circuit shown here to reduce the signal level and isolate the microphone circuit from the soundcard. (Multimode communication processors have this attenuator circuitry built in.)

T1—1:1 isolation transformer (RadioShack 273-1374)
R1—47 kΩ potentiometer (RadioShack 271-283)

The popular ICOM IC-706 transceiver only allows you to select narrow IF filters when you are in the CW or RTTY modes. But what happens when you're operating modes such as PSK31, Clover or PACTOR II that require SSB transmission?

To use narrow filters in the SSB mode, you have to fool the IC-706 into "thinking" that you're selecting a narrow SSB IF filter when you're really switching in a narrower 500 or 250-Hz filter.

With your 706 off, push the **LOCK** and **POWER** buttons together to access the setup menu. Let's assume your 500-Hz filter is installed in filter slot number 1. Select menu item 19 (OPT. FIL 1) and select the filter FL-223. You are effectively telling the radio that you have installed the FL-223, a 1.8 kHz filter, in that position. Of course, that's a lie! When you are operating in the USB mode and press the **FIL** (filter button), the IC-706 will activate *whatever filter is in that slot*. In this case, it will be a 500-Hz filter.

Still with me?

Once you've activated the narrower filter you'll need to adjust your **IF SHIFT** control to place the PSK31 receive signal within the narrower IF passband. If you're using the PSK31 software center frequency default of 1000 Hz, try setting the **IF SHIFT** control to the 1 o'clock position.

Similar techniques may also work with other ICOM transceivers, as well as Yaesu or Kenwood. Consult your manual.

print. There is a "gotcha" waiting for the unwary! Many transceivers are designed to allow you to select these narrow IF filters *only* when you have the radio in the "RTTY" (sometimes referred to as "data" or "FSK") or CW modes. That's fine if you are operating PACTOR, AMTOR or RTTY, but other modes such as PSK31, Clover and PACTOR II require the radio to operate in SSB. Many transceivers will *not* allow you to choose narrow IF filtering when you select SSB.

Does this mean that you should avoid buying transceivers that do not offer total IF filtering flexibility? Not at all. As I've already said, DSP audio filtering can do an adequate job, and some radios can even be "tricked" into selecting narrow IF filtering in the SSB mode (see the sidebar, "'Tricking' the ICOM IC-706"). The decision really depends on the application. If you want to indulge in digital contesting at a highly competitive level, it probably pays to invest in a radio that gives you complete freedom to select any IF filter in any operating mode. Otherwise, IF filtering is not a critical issue.

Chapter THREE

RTTY

As we discussed in Chapter 1, RTTY is the "old man" of the HF digital modes. Despite more than five decades of amateur use it remains a popular choice. There are several reasons for this:

(1) RTTY hardware is easy to set up. All it takes is one of the simple modems described in Chapter 2, or a PC equipped with a soundcard. You can transmit RTTY with *any* SSB transceiver—new or old.

(2) RTTY is easy to operate. Because RTTY is not a "handshaking" mode, you do not set up a link with elaborately timed protocols. Enjoying a RTTY conversation is a matter of just typing and transmitting. RTTY lends itself well to roundtable discussions, nets, contests and DX pileups where 100% error-free text is not required.

If you are new to HF digital, RTTY is a good mode to start. As with most HF digital modes, the majority of the activity is on 20 meters (see the frequency list in Chapter 1). You'll also hear RTTY occasionally on 40 and 15 meters. Of course, during contests you'll hear RTTY on almost every band!

MARK AND SPACE

Each character in the Baudot RTTY code is composed of five bits (see the sidebar, "The Baudot Code"). In amateur RTTY communication a "1" bit is usually represented by a 2125-Hz tone and is known as a *mark*. A "0" bit is represented by a 2295-Hz tone called a *space*. There is also a start pulse at the beginning of the bit string and a stop pulse at the end (see **Figure 3-1**). The data is commonly sent at a rate of 60 WPM, or 45 baud.

The Baudot Code

By Bill Henry, K9GWT

One of the first codes used with mechanical printing machines used a total of five data pulses to represent the alphabet, numerals and symbols. This code is commonly called the Baudot or Murray telegraph code after the work done by these two pioneers. Although commonly called the Baudot code in the United States, a similar code is usually called the Murray code in other parts of the world and is formally defined as the International Telegraphic No. 2 Baudot Code in part 97.69 of the FCC Rules. This standard defines the codes for letters, numbers, and the slant or fraction bar but allows variations in the choice of code combinations for punctuation. US amateurs have generally adopted a version of the so-called "Military Standard" code arrangement for punctuation, due largely to the ready availability of military surplus machines in the post-1945 years. Amateurs in other countries (particularly in Europe) have standardized on the CCITT No. 2 code arrangement that is similar to the U.S. standard, but has minor symbol and code arrangement differences.

Since each of the five data pulses can be in either a mark or space condition (two possible states per pulses), a total of $2 \times 2 \times 2 \times 2 \times 2 = 2$ to the fifth = 32 different codes combinations are possible. Since it is necessary to provide transmissions of all 26 letters, 10 numerals, plus punctuation, the 32 code combinations are not sufficient. This problem is solved by using the code twice: once in the Letters (LTRS) case and again in the Figures (FIGS) case. Two special characters, LTRS and FIGS, are used to indicate to the computer or processor whether the following characters will be of the letters or figures case.

The Baudot code has seen extensive commercial use throughout the world and is still actively utilized for international wire communications, press and weather. Due to the ready availability of Baudot mechanical equipment, this code will continue to be quite popular among radio amateurs. However, the lack of code space for control, extended punctuation, or lower case letters is a severe limitation of the 5-unit Baudot code. These limitations are particularly inconvenient in computer applications.

Table 3-1 depicts the Baudot Code Set (ITA#2). The leftmost bit is the Most Significant Bit (MSB), transmitted last. The rightmost bit is the Least Significant Bit (LSB), transmitted first. The associated letters and figures (case) characters are also listed, along with the hexadecimal representation of the characters.

Table 3-1
The Baudot Code

Bits	LTRS	FIGS	HEX
00011	A	-	03
11001	B	?	19
01110	C	:	0E
01001	D	$	09
00001	E	3	01
01101	F	!	0D
11010	G	&	1A
10100	H	STOP	14
00110	I	8	06
01011	J	'	0B
01111	K	(	0F
10010	L	)	12
11100	M	.	1C
01100	N	,	0C
11000	O	9	18
10110	P	0	16
10111	Q	1	17
01010	R	4	0A
00101	S	BELL	05
10000	T	5	10
00111	U	7	07
11110	V	;	1E
10011	W	2	13
11101	X	/	1D
10101	Y	6	15
10001	Z	"	11
00000	n/a	n/a	00
01000	CR	CR	08
00010	LF	LF	02
00100	SP	SP	04
11111	LTRS	LTRS	1F
11011	FIGS	FIGS	1B

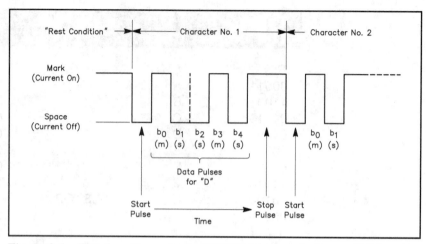

Figure 3-1—This is a diagram of a RTTY signal as the letter "D" is being sent. A start pulse begins the character, followed by the five bits (b0-b4) that define it. A stop pulse signals the end of the character.

Shift

Grab your calculator and do a bit of subtraction:
2295 Hz – 2125 Hz = 170 Hz

In the answer shown above, 170 Hz is the difference or *shift* between the mark and space frequencies. The Amateur Radio RTTY standard is to use either a 170- or 200-Hz shift. Multimode communications processors often use a 200-Hz shift with the mark tone at 2110 Hz and the space tone at 2310 Hz. RTTY soundcard software and transceivers running in the FSK mode tend to default to the 170-Hz shift. Fortunately, both shifts share the same center frequency (2210 Hz) which makes them *generally* compatible. This means that you don't usually need to worry whether the station you are attempting to contact is using a 170- or 200-Hz shift. In some weak-signal conditions, however, you may find that you need to tune carefully so that your shift "straddles" that of the other stations.

TUNING AND DECODING THE RTTY SIGNAL

Every RTTY decoder, whether it works in software or hardware, incorporates a set of mark and space audio filters. You can think of these filters as dual windows that only open for tones that are at the correct mark and space frequencies, and separated by the proper shift. The mark and space filtering circuitry detects and decodes the tones into digital 1s and 0s, which is exactly what your computer or processor needs to provide text on your screen. The more sensitive and selective your mark/space detectors, the better your RTTY performance, especially as it involves your ability to copy weak signals through interference.

It's easy to understand why tuning a RTTY signal (and just about any other data signal) is so critical—and why a good tuning indicator is one of your best HF digital tools. Whenever you stumble upon a RTTY signal, you must quickly tune your receiver until its mark and space tones fall within the "skirts" of the filters and are detected. It's possible to do this by ear once you become accustomed to the sound of a properly tuned RTTY transmission, but few of us have the necessary patience! Instead we rely on visual indicators to guide us.

The old method of tuning was to use an oscilloscope. You tweaked the signal until two ellipses representing the mark and space tones crossed in the middle of the screen. This was known as the *crossed bananas* display (see **Figure 3-2**). Some modern software and hardware tuning indicators pay homage to the crossed bananas by providing a display in the shape of a plus (+) sign, or cross. In hardware such as the HAL Communications DXP38 multimode processor, for example, you tune until the + display flashes in sync with the RTTY signal. The Kantronics, MFJ and Timewave processors use a "bouncing" LED bargraph, but the idea is fundamentally the same.

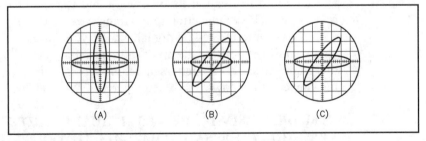

Figure 3-2—This is the classic crossed bananas RTTY tuning display. Pattern A indicates that the signal has been tuned corrected. At B the receiver is slightly off frequency, while C indicates that the transmitting station is using a shift that differs from your processor or modem setting. Although hardly any RTTY operators use oscilloscope tuning today, modern tuning indicators still rely on the same principle. Some even attempt to emulate the crossed bananas display.

The software tuning indicator provided by *WriteLog*, a RTTY contest software package (see Chapter 11).

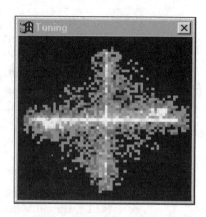

CONVERSING WITH RTTY

As with so many aspects of Amateur Radio, I highly recommend that you begin by listening. Tune between 14.070 and 14.099 MHz and listen for the long, continuous *blee-blee-blee-blee* signals of RTTY. (If you hear chirping, it isn't RTTY!)

Make sure your transceiver is set for lower sideband (LSB). That is the RTTY convention. The exception is when you are operating your rig in the "FSK" mode (sometimes labeled "DATA" or "RTTY"). As we discussed in Chapter 2, many hams prefer operating this way because it allows them to use narrow IF filters to reduce interference. When you operate in FSK, your modem, soundcard or processor is *not* generating the mark and space tones. They're merely sending data pulses to the radio and the radio is creating its own mark and space signals. (This often requires a special connection to the transceiver. Consult your manual.)

As you tune the RTTY signal, watch your tuning indicator. Tune slowly until you see that the mark and space tones are being decoded. At this point you should see letters marching across your screen. Notice how the conversation flows just like a voice or CW ragchew.

KF6I DE WB8IMY . . . YES, I HEARD FROM SAM JUST YESTERDAY. HE SAID THAT HIS TOWER PROJECT WAS ALMOST FINISHED. KF6I DE WB8IMY K

If you want to call CQ, the procedure is simple. Some programs have "canned" CQ messages that you can customize with your call sign. Others allow you to type off the air, filling a buffer with your CQ and storing it temporarily until you're ready to go.

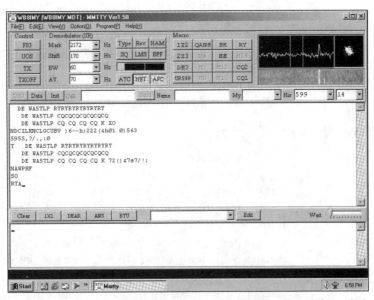

There are many ways to call CQ on RTTY. Here is a CQ copied from WA5TLP using *MMTTY* software on 20 meters. The number/character gibberish is caused by noise between transmissions.

Let's assume that you have a CQ stored in your buffer right now. Press the key that puts your rig in the transmit mode. Now tap the key that spills the contents of your buffer into the modem. If you're monitoring your own signal you'll hear the delightful chatter of RTTY and see something like this on your screen:

CQ CQ CQ CQ CQ CQ WB8IMY WB8IMY WB8IMY
CQ CQ CQ CQ CQ CQ WB8IMY WB8IMY WB8IMY K K

Now jump back to receive. That's all there is to it!

Notice how my CQ is short and to the point. Did you also notice that I repeated everything several times? *Remember that RTTY lacks error detection.* If you want to make certain that the other station copied what you sent, it helps to repeat it. You'll see this often in contest exchanges.

For example:

N6ATQ DE N1RL . . . UR 549 549. STATE IS CT CT. DE N1RL K

On the other hand, if you know that the other station is copying you well, there is no need to repeat information.

The Mystery Signal

What happens if you stumble across a RTTY signal that you cannot copy? The signal is strong enough, and you seem to be doing all the right things, but you see gibberish on your screen or nothing at all. Why?

It looks like you need to do a little detective work. Check the following:

• Is the signal "upside down"? The RF frequency of the mark signal is usually higher than the RF frequency of the space signal, but there is no law that dictates this standard. With most processors and software you can flip the relationship with the push of a button or the click of a mouse.

• Is the station running in upper sideband (USB) rather than lower sideband (LSB)?

• Are the operators really using signals that have 170- or 200-Hz shifts at 45 baud? Some RTTY operators might choose to run at 75 baud when sending lengthy text. Others may choose to use oddball shifts.

CHASING RTTY DX

RTTY is still the most widely used mode for HF digital DXing. You can work 100 countries to earn your RTTY DX Century Club (DXCC) award, and even climb the ladder to the stratospheric heights of RTTY DXCC Honor Roll. RTTY DXing is competitive and highly rewarding.

Like any other form of DXing, the quest for RTTY DX demands patience and skill. When a DXpedition is on the air with RTTY from a rare DXCC entity, your signal will be in competition with thousands of other HF digital operators who want to work the station as badly as you! Sometimes pure luck is the winning factor, but there are several tricks of the trade that you can use to tweak the odds in your favor.

Don't Call...Yet

Let's say that you're tuning through the HF digital subbands one day and you stumble across a screaming mass of RTTY signals. On your computer screen you see that everyone seems to be frantically calling a

DX station. Oh, boy! It's a pileup!

You can't actually hear the DX station that has everyone so excited, but what the heck, you'll activate your transceiver and throw your call sign into the fray, right? *Wrong!*

Never transmit even a microwatt of RF until you can copy the DX station. Tossing your call sign in blindly is pointless and will only add to the pandemonium. Instead, take a deep breath and wait. When the calls subside, can you see text from the DX station on your screen? If not, the station is probably too weak for you to work (don't even bother), or he may be working "split." More about that in a moment.

If you can copy the DX station, watch the exchange carefully. Is he calling for certain stations only? In other words, is he sending instructions such as "North America only"? Calling in direct violation of the DX station's instructions is a good way to get yourself blacklisted in his log. (No QSL card for you—ever!) Does he just want signal reports, or is he in the mood for brief chats? Most DX stations simply want "599" and possibly your location—period. Don't give them more than they are asking for. (A DX RTTY station on a rare island doesn't care what kind weather you are experiencing at the moment.)

Working the Split

When DX RTTY pileups threaten to spin out of control, many DX operators will resort to working "split." In this case, "split" means split frequency. The DX station will transmit on one frequency while listening for calls on another frequency (or range of frequencies).

A good DX operator will announce the fact that he is working split with almost every exchange. That's why it is so important to listen to a pileup before you throw yourself into the middle. If you tune into a pileup and cannot hear the DX station, tune below the pileup and see if you copy the DX station there. If his signal is strong enough, he shouldn't be hard to find if he is working split. His signal will seem to be by itself, answering calls that you cannot hear. This is a major clue that a split operation is taking place. Watch for copy such as...

CQ DX DE FO0AAA, UP 10 (He is listening up 10 kHz)

CQ DX DE FO0AAA, 14085-14090 (He is listening between 14085 and 14090 kHz)

Whatever you do, *never call a split DX operation on the station's transmitting frequency.* Your screen will quickly fill with rude comments from others who are listening. Instead, put your transceiver into the split-frequency mode (better drag out your rig manual!). Set your radio to receive on the DX station's transmitting frequency and transmit on his listening frequency. If the DX station is listening through a range of frequencies, you'll need to select a spot where you think you'll be heard. Change your transmit frequency if this particular "fishing spot" doesn't seem to be working.

Short and Sweet

When you've done your listening homework and you're ready for battle, by all means fire at will. Wait until the DX station finishes an exchange. He'll signal that he is ready for another call by sending "QRZ?" or something similar. When it's time to transmit, make it short and to the point, like this...

WB8IMY WB8IMY WB8IMY K

Listen again. Has he responded to anyone yet? If the answer is "no" and other stations are still calling, the DX is probably trying to sort out the alphabet soup of confusion on his screen. Fire again!

WB8IMY WB8IMY WB8IMY KK

You might make it through at just the right moment when other signals subside briefly, or when the ionosphere gives you an unexpected boost. But if no one is calling, or if you hear the DX station calling someone else, *stop*. You lost this round, so give the lucky winner his chance to be heard. Your next opportunity will be coming up shortly.

Did He Call You?

Watch your screen carefully. If the DX station only copied a fragment of your call sign, he might send something like...

IMY IMY AGAIN??

In this instance he copied only the last three letters of my call sign ("IMY"). Fading and flutter can make RTTY signals difficult to copy clearly. See **Figure 3-3**. The best thing to do is reply right away, sending your call sign three times just like before.

If luck is on your side, you'll see...

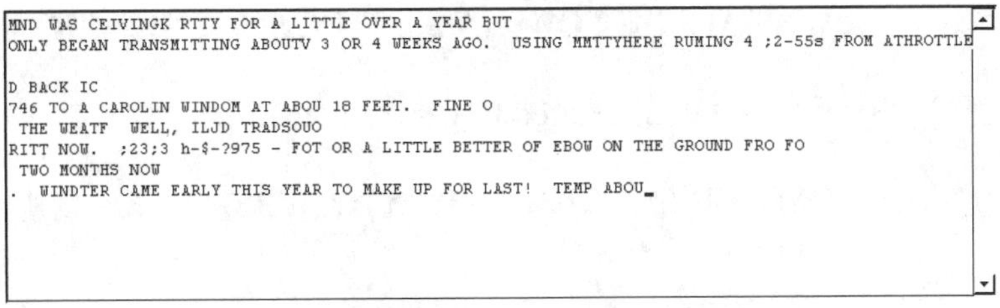

```
MND WAS CEIVINGK RTTY FOR A LITTLE OVER A YEAR BUT
ONLY BEGAN TRANSMITTING ABOUTV 3 OR 4 WEEKS AGO.  USING MMTTYHERE RUMING 4 ;2-55s FROM ATHROTTLE

D BACK IC
746 TO A CAROLIN WINDOM AT ABOU 18 FEET.  FINE O
 THE WEATF  WELL, ILJD TRADSOUO
RITT NOW.  ;23;3 h-$-?975 - FOT OR A LITTLE BETTER OF EBOW ON THE GROUND FRO FO
 TWO MONTHS NOW
 .  WINDTER CAME EARLY THIS YEAR TO MAKE UP FOR LAST!  TEMP ABOU_
```

Figure 3-3—Here is an actual on-air example of how fading can
distort a RTTY DX signal. Notice the missing or incorrect letters. This
example represents about 50-60% copy, which is probably sufficient
to at least make a brief DX contact.

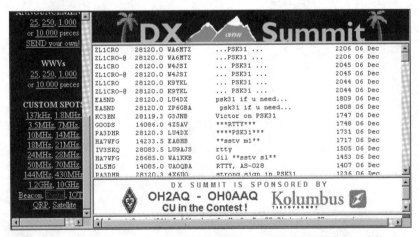

Figure 3-4—This is the OH9W/OH2AQ Radio Club "DX Summit"
WebCluster at *oh2aq.kolumbus.com/dxs/*. I've just requested a list of
digital DX reports.

WB8IMY DE FOØAAA . . . TNX. 599. QSL? KK

And with excited fingers you answer...

QSL. UR 599. TNX AND 73 DE WB8IMY K

TRACKING THE DX

The most direct method for tracking RTTY DX is to tune through the bands and see who is on the air. With practice you'll learn when particular areas of the world are open to you. For example, 10 meters tends to open from the early afternoon into early evening while 40 meters is only open for DX at night. Twenty meters can provide DX at almost any time, which is why it is the most popular RTTY band.

Another method is to check the online DX WebClusters (see **Figure 3-4**). These sites collect DX information from operators throughout the world. Don't depend too heavily on these sites, though. Many DX "sightings" go unreported. There is no substitute for good, old-fashioned listening.

Chapter FOUR

PSK31

PSK31 was the brainchild of Peter Martinez, G3PLX, the father of AMTOR (see Chapter 1). He wanted to create a mode that was as easy to use as RTTY, yet much more robust in terms of weak-signal performance. Another important criterion was bandwidth. The HF digital subbands are narrow and tend to become crowded in a hurry (particularly during contests). Peter wanted to design a mode that would do all of its tricks within a very narrow bandwidth.

Experiments with PSK31 began in the mid-'90s. Amateurs at that time used *DOS* software and DSP development platforms such as the Texas Instruments TMS320C50DSK, Analog Devices ADSP21061 or Motorola DSP56002EVM. But in early 1999 Peter Martinez unveiled the first PSK31 software for *Windows*. The program required only a common 16-bit soundcard functioning as the analog-to-digital converter (and vice versa). With the large number of hams owning PCs equipped with soundcards, the popularity of PSK31 exploded!

SO WHAT IS PSK31?

First, let's dissect the name. The "PSK" stands for Phase Shift Keying, the modulation method that is used to generate the signal; "31" is the bit rate. Technically speaking, the bit rate is really 31.25, but "PSK31.25" isn't nearly as catchy.

Think of Morse code for a moment. It is a simple binary code expressed by short signal pulses (*dits*) and longer signal pulses (*dahs*). By combining strings of dits and dahs, we can communicate the entire

English alphabet along with numbers and punctuation. Morse uses gaps of specific lengths to separate individual characters and words. Even beginners quickly learn to recognize these gaps—they don't need special signals to tell them that one character or word has ended and another is about to begin.

When it comes to RTTY we're still dealing with binary data (dits and dahs, if you will), but instead of on/off keying, we send the information by shifting frequencies. This is known as Frequency Shift Keying or *FSK*. One frequency represents a *mark* (1) and another represents a *space* (0). If you put enough mark and space signals together in proper order according to the RTTY code, you can send letters, numbers and a limited amount of punctuation.

The RTTY code shuffles various combinations of five bits to represent each character. For example, the letter A is expressed as 00011. To separate the individual characters RTTY must also add "start" and "stop" pulses (see Chapter 3).

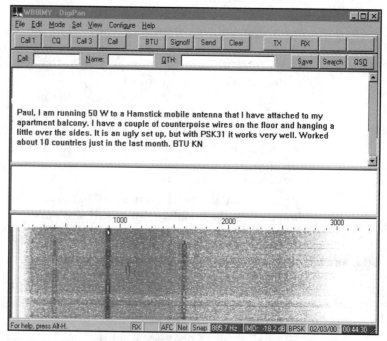

Digipan pioneered the panoramic method of receiving PSK31 signals. Each line represents a PSK31 signal. You can copy individual signals by placing your mouse cursor on the line of your choice, then right clicking your mouse.

Table 4-1
The PSK31 Varicode

The codes are transmitted left bit first, with "0" representing a phase reversal on BPSK and "1" representing a steady carrier. A minimum of two zeros is inserted between characters. Some implementations may not handle all the codes below 32.

ASCII*	Varicode	ASCII*	Varicode	ASCII*	Varicode
0 (NUL)	1010101011	.	1010111	\	111101111
1 (SOH)	1011011011	/	110101111	]	111111011
2 (STX)	1011101101	0	10110111	^	1010111111
3 (ETX)	1101110111	1	10111101	_	101101101
4 (EOT)	1011101011	2	11101101	'	1011011111
5 (ENQ)	1101011111	3	11111111	a	1011
6 (ACK)	1011101111	4	101110111	b	1011111
7 (BEL)	1011111101	5	101011011	c	101111
8 (BS)	1011111111	6	101101011	d	101101
9 (HT)	11101111	7	110101101	e	11
10 (LF)	11101	8	110101011	f	111101
11 (VT)	1101101111	9	110101111	g	1011011
12 (FF)	1011011101	:	11110101	h	101011
13 (CR)	11111	;	110111101	i	1101
14 (SO)	1101110101	<	111101101	j	111101011
15 (SI)	1110101011	=	1010101	k	10111111
16 (DLE)	1011110111	>	111010111	l	11011
17 (DC1)	1011110101	?	1010101111	m	111011
18 (DC2)	1110101101	@	1010111101	n	1111
19 (DC3)	1110101111	A	1111101	o	111
20 (DC4)	1101011011	B	11101011	p	111111
21 (NAK)	1101101011	C	10101101	q	110111111
22 (SYN)	1101101101	D	10110101	r	10101
23 (ETB)	1101010111	E	1110111	s	10111
24 (CAN)	1101111011	F	11011011	t	101
25 (EM)	1101111101	G	11111101	u	110111
26 (SUB)	1110110111	H	101010101	v	1111011
27 (ESC)	1101010101	I	1111111	w	1101011
28 (FS)	1101011101	J	111111101	x	11011111
29 (GS)	1110101011	K	101111101	y	1011101
30 (RS)	1011111011	L	11010111	z	111010101
31 (US)	1101111111	M	10111011	{	1010110111
32 (SP)	1	N	11011101			110111011
!	111111111	O	10101011	}	1010110101
"	101011111	P	11010101	~	1011010111
#	111110101	Q	111011101	127	1110110101
$	111011011	R	10101111		
%	1011010101	S	1101111		
&	1010111011	T	1101101		
'	101111111	U	101010111		
(	11111011	V	110110101		
)	11110111	W	101011101		
*	101101111	X	101110101		
+	111011111	Y	101111011		
,	1110101	Z	101010101101		
-	110101	[	111110111		

*ASCII characters 0 through 32 are control codes. Their abbreviations are shown here in parentheses. For the meanings of the abbreviations, refer to any recent *ARRL Handbook*.

For PSK31 Peter devised a new code that combines the best of RTTY and Morse. He christened his creation the *Varicode* because a varying number of bits are used for each character (see **Table 4-1**). Building on the example of Morse, Peter allocated the shorter codes to the letters that appeared most often in standard English text. The idea was to send the least number of bits possible during a given transmission. For example:

E is a very popular letter, so it gets a Varicode of 11

Z sees relatively little use, so its Varicode becomes 111010101

As with RTTY, however, we still need a way to signal the gaps between characters. The Varicode does this by using "00" to represent a gap. The Varicode is carefully structured so that two zeros never appear together in any of the combinations of 1s and 0s that make up the characters.

But how would the average ham generate a BPSK signal and transmit Varicode over the airwaves? Peter's answer was to use the DSP capabilities of the computer soundcard to create an audio signal that shifted its phase 180° in sync with the 31.25 bit-per-second data stream. In Peter's scheme, a 0 bit in the data stream generates an audio phase shift, but a 1 does not. The technique of using phase shifts (and the lack thereof) to represent binary data is known as Binary Phase-Shift Keying, or *BPSK*. If you apply a BPSK audio signal to an SSB transceiver, you end up with BPSK modulated RF. At this data rate the resulting PSK31 RF signal is only 31.25 Hz wide, which is actually narrower than the average CW signal!

Concentrating your RF into a narrow bandwidth does wonders for reception, as any CW operator will tell you. But when you're trying to receive a BPSK-modulated signal it is easier to recognize the phase transitions—even when they are deep in the noise—if your computer knows when to expect them. To accomplish this, the receiving station must synchronize with the transmitting station. Once they are in sync, the software at the receiving station "knows" when to look for data in the receiver's audio output. Every PSK31 transmission begins with a short "idle" string of 0s. This allows the receive software to get into sync right away. Thanks to the structure of the Varicode, however, the phase transitions are also mathematically predictable, so much so that the PSK31 software can quickly synchronize itself when you tune in during the middle of a transmission, or after you momentarily lose the signal.

The combination of narrow bandwidth, an efficient DSP algorithm and synchronized sampling creates a mode that can be received at very low signal levels. PSK31 rivals the weak-signal performance of CW and it is a *vast* improvement over RTTY.

Its terrific performance notwithstanding, PSK31 will not always provide 100% copy; it is as vulnerable to interference as any digital mode. And there are times, during a geomagnetic storm, for example, when ionospheric propagation will exhibit poor frequency stability. (When you are trying to receive a narrow-bandwidth, phase-shifting signal, frequency stability is very important.) This effect is almost always confined to the polar regions and it shows up as very rapid flutter, which is deadly to PSK31. The good news is that these events are usually short-lived.

From BPSK to QPSK

Many people urged Peter to add some form of error correction to PSK31, but he initially resisted the idea because most error-correction schemes rely on transmitting redundant data bits. Adding more bits while still maintaining the desired throughput required a doubling of the data rate. If you double the BPSK data rate, the bandwidth doubles. As the bandwidth increases, the signal-to-noise ratio deteriorates and you get more errors. It's a sticky digital dilemma. How do you expand the information capacity of a BPSK channel without significantly increasing its bandwidth?

Peter finally found the answer by adding a second BPSK carrier at the transmitter with a 90° phase difference and a second demodulator at the receiver. Peter calls this quadrature polarity reversed keying, but it is better known as quaternary phase-shift keying or *QPSK*.

Splitting the transmitter power between two channels results in a 3-dB signal-to-noise penalty, but this is the same penalty you'd suffer if you doubled the bandwidth. Now that we have another channel to carry the redundant bits, we can use a *convolutional encoder* to generate one of four different phase shifts that correspond to patterns of five successive data bits. On the receiving end we have a *Viterbi* decoder playing a very sophisticated guessing game. See the sidebar "The Viterbi Competition and QPSK."

Operating PSK31 in the QPSK mode will give you 100% copy under most conditions, but there is a catch. Tuning is twice as critical with QPSK as it is with BPSK. You have to tune the received signal within an accuracy of less than 4 Hz for the Viterbi decoder to detect the phase shifts and do its job. Obviously, both stations must be using very stable transceivers.

PSK31 SOFTWARE

The first step in setting up a PSK31 station is to jump onto the Web

THE VITERBI COMPETITION AND QPSK

The Viterbi is not so much a decoder as a whole family of encoders. Each one makes a different "guess" at what the last five transmitted QPSK data bits might have been. There are 32 different patterns of five bits and thus 32 encoders. At each step the phase-shift value predicted by the bit-pattern guess from each encoder is compared with the actual received phase-shift value, and the 32 encoders are given "marks out of 10" for accuracy. Just like in a knockout competition, the worst 16 are eliminated and the best 16 go on to the next round, taking their previous scores with them. Each surviving encoder then gives birth to 'children,' one guessing that the next transmitted bit will be a 0 and the other guessing that the next transmitted bit will be a 1. They all do their encoding to guess what the next phase shift will be, and are given marks out of 10 again that are added on to their earlier scores. The worst 16 encoders are killed off again and the cycle repeats.

It's a bit like Darwin's theory of evolution, and eventually all the descendants of the encoders that made the right guesses earlier will be among the survivors and will all carry the same "ancestral genes." We therefore just keep a record of the family tree (the bit-guess sequence) of each survivor, and can trace back to find the transmitted bit stream, although we have to wait at least five generations (bit periods) before all survivors have the same great great grandmother (who guessed right five bits ago). The whole point is that because the scoring system is based on the running total, the decoder always gives the most accurate guess—*even if the received bit pattern is corrupted.* In other words, the Viterbi decoder corrects errors.—*Peter Martinez, G3PLX*

and download the software you need, according to the type of computer system you are using. As this book went to press there were PSK31 programs available for *Windows* (98, 95 and 3.1), *Linux*, *DOS* and the Mac. The Resources section of this book has lists of Web addresses for software downloading.

Once you have the software safely tucked away on your hard drive, install it and read the "Help" files. Every PSK31 program has different features—too many to cover in this chapter. Besides, the features will no doubt change substantially in the months and years that follow the publication of this book. The documentation that comes with your software is the best reference.

PSK31: A TESTIMONIAL

At the time this book went to press, I had been a PSK31 user for eight months. I've been running PSK31 using my ICOM IC-706 MkII transceiver and my end-fed wire long-wire antenna. To say the results have been impressive is an understatement!

My first PSK31 contact was with K8SRB on 40 meters. Stan was only running 25 W to a G5RV dipole antenna, but the text on my screen was virtually error-free despite the high noise levels. We chatted for about 45 minutes and Stan passed along a wealth of PSK31 tips. For example, he demonstrated the need to keep your transmit audio in line by momentarily overdriving his rig so I could hear the difference and see it on the waterfall display. (The splatter appeared as vertical lines to the left and right of the center position.) Stan also clued me in on the importance of checking the NET box so that our transmit and receive frequencies would track each other. As Stan put it, "If we must drift, let's drift together!"

A few days later I had the spooky experience of being able to copy text from a signal I could not hear. I was tuning across the 20-meter digital subband when I saw a faint, ghostly trace on the waterfall display. I watched the display and tuned in the signal, but I could hear nothing recognizable in the noise. There may have been a warbling tone present, but I couldn't be sure. Suddenly the text began to print a CQ from WL7VO in Chicken, Alaska near Fairbanks. I answered and was astonished to see his reply. The contact didn't last long because the band was dying fast, but it was captivating nonetheless. If we had been using RTTY a QSO would have been utterly impossible; even with CW it would have been a challenge.

Since that time I've logged dozens of PSK31 contacts, including a fair amount of DX. The mode is catching on quickly in Europe, Australia and Japan. In fact, the activity level has risen to the point where you can find PSK31 signals on the air just about any time 20 meters is open. I've also been plumbing the depths of 80 and 160 meters, looking for PSK31 signals amidst the noise. The few contacts I've had on 160 have proven the power of PSK31. Despite Mother Nature's static crescendo I was able to copy readable text.—*Steve Ford, WB8IMY*

The Panoramic Approach

All of the PSK31 programs are good; there isn't a bad apple in the bunch. But when I was introduced to *DigiPan*, it was love at first sight.

One of the early bugaboos with PSK31 had to do with tuning. Most PSK31 programs required you to tune your radio carefully, preferably in 1-Hz increments. In the case of the original G3PLX software, for example, the narrow PSK31 signal would appear as a white trace on a thin *waterfall* display. Your goal was to bring the white trace directly into the center of the display, then tweak a bit more until the phase indicator in the circle above the waterfall was more-or-less vertical (or in the shape of a flashing cross if you were tuning at QPSK signal). Regardless of the software, PSK31 tuning required practice. You had to learn to recognize the sight and sound of your target signal. With the weak warbling of PSK31, that wasn't always easy to do. And if your radio didn't tune in 1-Hz increments, the receiving task became even more difficult.

Nick Fedoseev, UT2UZ and Skip Teller, KH6TY, designed a solution and called it *DigiPan*. The "pan" in *DigiPan* stands for "panoramic"—a complete departure from the way most PSK31 programs work. With *DigiPan* the idea is to eliminate tedious tuning by detecting and displaying not just one signal, but entire *groups* of signals.

If you are operating your transceiver in SSB without using narrow IF or audio-frequency filtering, the bandwidth of the receive audio that you're dumping to your soundcard is about 2000 to 3000 Hz. With a bandwidth of only about 31 Hz, a lot of PSK31 signals can squeeze into that spectrum. *DigiPan* acts like an audio spectrum analyzer, sweeping through the received audio from 100 to 3000 Hz and showing you the results in a large waterfall display that continuously scrolls from top to bottom. What you see on your monitor are vertical lines of various colors that indicate every signal that *DigiPan* can detect. Bright yellow lines represent strong signals while blue lines indicate weaker signals.

The beauty of *DigiPan* is that you do not have to tune your radio to monitor any of the signals you see in the waterfall. You simply move your mouse cursor to the signal of your choice and click. A black diamond appears on the trace and *DigiPan* begins displaying text. You can hop from one signal to another in less than a second merely by clicking your mouse! If you discover someone calling CQ and you want to answer, click on the transmit button and away you go—no radio adjustments necessary. (And like the original PSK31 software, *DigiPan* automatically corrects for frequency drift.)

DON'T OVERDRIVE YOUR TRANSCEIVER!

When you're setting up your rig for your first PSK31 transmission, the temptation is to adjust the output settings for "maximum smoke." This can be a serious mistake because overdriving your transceiver in PSK31 can result in a horrendous amount of splatter, which will suddenly make your PSK31 signal much wider than 31 Hz—and make you highly unpopular with operators on adjacent frequencies.

As you increase the transmit audio output from your soundcard or multimode processor, watch the ALC indicator on your transceiver. The ALC is the automatic level control that governs the audio drive level. When you see your ALC display indicating that audio limiting is taking place, you are feeding too much audio to the transceiver. The goal is to achieve the desired RF output with little or no activation of the ALC.

Unfortunately, monitoring the ALC by itself is not always a sure bet. Many radios can be driven to full output without budging the ALC meter. You'd think that it would be smooth sailing from there, but a number of rigs become decidedly nonlinear when asked to provide SSB output beyond a certain level (sometimes this nonlinearity can begin at the 50% output level). We can ignore the linearity issue to a certain extent with an SSB voice signal, but not with PSK31 because the immediate result, once again, is splatter.

So how can you tell if your PSK31 signal is really clean? Unless you have the means to monitor your RF output with an oscilloscope, the only way to check your signal is to ask someone to give you an evaluation on the air. The PSK31 programs that use a waterfall audio spectrum display can easily detect "dirty" signals. The splatter appears as rows of lines extending to the right and left of your primary signal. (Overdriven PSK31 signals may also have a harsh, clicking sound.)

If you are told that you are splattering, ask the other station to observe your signal as you slowly decrease the audio level from the soundcard or processor. When you reach the point where the splatter disappears, you're all set. Don't worry if you discover that you can only generate a clean signal at, say, 50 W output. With PSK31 the performance differential between 50 W and 100 W is inconsequential.
—*Steve Ford, WB8IMY*

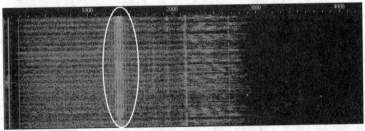

An overmodulated PSK31 signal. Notice the lines to the left and right of the primary signal. This is actually a mild example of overmodulation. Serious offenders will spread over much more spectrum!

To transmit Varicode at a reasonable typing speed of about 50 words per minute needs a bit-rate of about 32 bps. In theory, we only need a bandwidth of 31.25 Hz to send this as binary data, and the frequency stability that this implies can be achieved with modern radio equipment on HF.

The method chosen was first used on the amateur bands, to my knowledge, by SP9VRC. Instead of frequency-shifting the carrier, which is wasteful of spectrum, or turning the carrier on and off, which is wasteful of transmitter power capability, the "dots" of the code are signaled by reversing the polarity of the carrier. You can think of this as equivalent to transposing the wires to your antenna feeder. This uses the transmitted signal more efficiently since we are comparing a positive signal before the reversal to a negative signal after it, rather than comparing the signal present in the dot to no-signal in the gap. But if we keyed the transmitter in this way at 31.25 baud, it would generate terrible key clicks, so we need to filter it.

If we take a string of dots in Morse code, and low-pass filter it to the theoretical minimum bandwidth, it will look the same as a carrier that is 100% amplitude-modulated by a sine wave at the dot rate. The spectrum is a central carrier and two sidebands at 6dB down on either side. A signal that is sending continuous reversals, filtered to the minimum bandwidth, is equivalent to a double-sideband suppressed-carrier emission, that is, to two tones either side of a suppressed carrier. The improvement in the performance of this polarity-reversal keying over on-off keying is thus equivalent to the textbook improvement in changing from amplitude-modulation telephony with full carrier to double-side-band with suppressed carrier. I have called this technique "polarity-reversal keying" so far, but everybody else calls it "binary phase-shift keying," or BPSK.

To generate BPSK in its simplest form, we could convert our data stream to levels of ± 1 V, for example, take it through a low-pass filter and feed it into a balanced modulator.

The other input to the balanced modulator is the desired carrier frequency. When sending continuous reversals, this looks like a 1 V (P-P) sine wave going into a DSB modulator, so the output is a pure two-tone signal. In practice we use a standard SSB transceiver and perform the modulation at audio frequencies or carry out the equivalent process in a DSP chip. We could signal logic zero by continuous carrier and signal logic one by a reversal, but I do it the other way round for reasons that will become clear shortly.

There are many ways to demodulate BPSK, but they all start

with a band-pass filter. For the speed chosen for PSK31, this filter can be as narrow as 31.25 Hz in theory. A brick-wall filter of precisely this width would be costly, however, not only in monetary terms but also in the delay time through the filter, and we want to avoid delays. A practical filter might be twice that width (62.5 Hz) at the 60-dB-down points with a delay-time of two bits (64 ms).

For the demodulation itself, since BPSK is equivalent to double sideband, the textbook method for demodulating DSB can be used. However, delaying the signal by one bit period and comparing it to the signal with no delay in a phase comparator can also demodulate it. The output is negative when the signal reverses polarity and positive when it doesn't.

We could extract the information from the demodulated signal by measuring the lengths of the "dots" and "dashes," as we do by ear with Morse code. It helps to pick the data out of the noise, however, if we know when to expect signal changes. We can easily transmit the data at an accurately timed rate, so it should be possible to predict when to sample the demodulator output. This process is known as synchronous reception, although the term "coherent" is sometimes wrongly used.

To synchronize the receiver to the transmitter, we can use the fact that a BPSK signal has an amplitude-modulation component. Although the modulation varies with the data pattern, it always contains a pure-tone component at the baud rate. This can be extracted using a narrow filter, a PLL or the DSP equivalent, and fed to the decoder to sample the demodulated data.

For the synchronization to work we need to make sure that there are no long gaps in the pattern of reversals. A completely steady carrier has no modulation, so we could never predict when the next reversal was due. Fortunately, Varicode is just what we need, provided we choose the logic levels so that zero corresponds to a reversal and one to a steady carrier. The idle signal of continuous zeros thus generates continuous reversals, giving us a strong 31.25-Hz modulation. Even with continuous keying, there will always be two reversals in the gaps between characters. The average number of reversals will therefore be more than two in every 6.5 bits, and there will never be more than 12 bits with no reversal at all. If we make sure that the transmission always starts with an idle period, then the timing will pull into sync quickly. By making the transmitter end a transmission with a "tail" of unmodulated carrier, it is then possible to use the presence or absence of reversals to squelch the decoder. Hence, the screen doesn't fill with noise when there is no signal.—*Peter Martinez, G3PLX*

The PSK20 transceiver kit by Dave Benson, NN1G, was an instant hit among PSK31 QRP enthusiasts. You'll find more information about the PSK20 online at the Small Wonder Labs Web site at: *www.smallwonderlabs.com*.

Another NN1G invention was the *Warbler*, a tiny 80-meter PSK31 transceiver that uses a direct conversion receiver. The kit was described in an article in the March 2001 issue of *QST*.

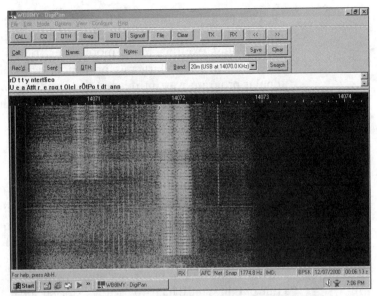

PACTOR interference, as seen with *DigiPan* software. In this example there is a very strong PACTOR signal (in the center of the waterfall display) and another weaker PACTOR signal nearby.

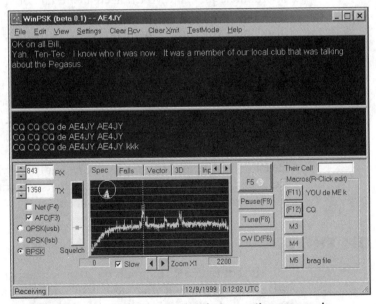

WinPSK is another PSK31 program that uses the panoramic approach.

This panoramic approach to digital signal communication is, in my opinion, one of the most important developments in the history of PSK31. It makes this exciting mode more "user friendly" and accessible to a larger audience and a larger assortment of radios. In recent years another popular program was introduced that used panoramic tuning: *WinPSK*. You can download both *DigiPan* and *WinPSK* on the Web for free. See the Resources section of this book.

TRANSCEIVER SETUP

Assuming that you have a reasonably stable HF SSB rig, you'll need to run shielded audio cables between your transceiver and your computer or processor as shown in Chapter 2. If your radio has an accessory jack that offers an audio line output, this is the preferred way to feed the receive audio. Connect one shielded cable between the radio line output and the soundcard or processor audio *input*. If your radio does not have a line output, you'll have to use the external speaker jack (see Chapter 2).

For transmit audio, use another shielded cable and connect it between your soundcard or processor audio output and the accessory audio *input* of your transceiver. You can also opt to route the transmit audio to your microphone jack, but you'll need an attenuator similar to the one shown in Chapter 2. If you use the accessory audio input, don't forget to disconnect your microphone before you go on the air. When you key the transceiver, the microphone may be "live," too!

And what about keying your transceiver? For multimode communications processor owners the solution is relatively easy. The processor provides a keying output that you can connect to the keying line at your transceiver (usually at the accessory jack).

For soundcard users there are two options: Use one of your PC's COM ports and an interface like the one shown in Chapter 2 to key your rig via the PTT line at your accessory jack. Or, simply switch on your transceiver's VOX and let it key the rig when it detects the transmit audio from the sound card. Once you have all of your cables in place, you're ready to go.

ON THE AIR

Most of the PSK31 signals on 20 meters are clustered around 14.070 MHz, start by parking your radio in the vicinity and booting up *DigiPan*. **Do not touch your rig's VFO again.** Just place your mouse cursor on one of the vertical signal lines and right click. That's all there is to it!

RF FEEDBACK AND SOUNDCARDS

One of the most common problems encountered when using a soundcard as a PSK31 modem is RF feedback. Soundcards are not designed for Amateur Radio, and can suffer from high levels of RF, usually getting into the card via the audio input and/or output cables from the transceiver. The RF causes nonlinear rectification in the soundcard circuitry, and the envelope of the rectified RF gets fed, as audio, to the transmitter again. This creates a loop which can result in a parasitic oscillation on the audio output of the soundcard, and hence also on the transmitter output. It usually takes the form of an extra tone that is exactly one-half of the wanted output audio. Thus, if the soundcard is outputting 1000Hz, the spurious is 500Hz and just sounds like an added richness to the sound. Alternatively the feedback may just create a jumble of noise.

The way to check for RF feedback is to listen to the soundcard output as you are transmitting, not by turning up the speakers (which may not be suffering) but by connecting headphones across the line to the transmitter. First listen with the transmitter off radiating, then see if there's any "added richness" in the sound as you switch the transmitter on and wind up the power. You may find, if there is RF feedback, that at a certain power level there is a sudden change in the sound as the parasitic oscillation starts.

Another way to detect RF feedback is if you have an old wavemeter with a headphone socket, intended originally for checking the audio of an AM transmitter. Alternatively you can make one yourself with just a pair of headphones and a single diode. The idea is simply to detect any amplitude modulation on the transmitter output. A clean transmission will give no sound at all (no amplitude modulation) when radiating a "tune" carrier, and a very low-frequency buzz (31.25Hz) when idling. RF feedback will give you a tone in the headphones that starts suddenly as you wind up the power past the critical parasitic-oscillation point. The cure is almost always to stop the RF getting down the audio cable from the transmitter to the sound card. Ferrite rings, with both the audio input and audio output wrapped together through the ring as many times as you can, will usually fix it. It's possible that the PTT cable from the transmitter to the COM port might need treating in the same way.—*Peter Martinez, G3PLX*

"MY PSK31 DOESN'T WORK!"

Problem: I know I'm getting receive audio from my radio to my sound card. I can even hear my radio's audio in my computer speakers. The PSK31 software, however, is dead as a doornail. There is nothing whatsoever in the tuning display.

Solution: Your receive audio is indeed reaching your computer, but it is not reaching the PSK31 software. This is a common glitch in *Windows 95* and *98* and it involves misadjustment of the sound card mixer. Follow these steps:

1. Double click on the little loudspeaker in the lower right corner of your *Windows* screen.
2. The **Volume Control** panel should appear. Click on **Options**, then **Properties**.
3. In the "**Adjust volume for...**" section, click on the circle labeled **Recording**, then click **OK**.
4. The mixer panel you see now is the **Recording Control**. Click on the little box to select **Line-In**, then move the slider all the way up.

   Note that some programs may disable this setting. Don't be surprised if you have to repeat these steps after running certain types of software.

Problem: People keep telling me that they are hearing strange noises on my PSK31 signal.

Solution: This is another common woe—and it is easy to fix. The strange noises are the cute little boops and beeps that you've asked *Windows* to play when you open a program, close a window and so on. The straightforward cure is to avoid opening windows or taking other actions that generate cute sounds while you are transmitting. You can also turn the offending sounds off by going to **Settings**, **Control Panel** and clicking on the **Sounds** icon. This will allow you to scroll through the list of sounds you've chosen. By changing the selection to "**None**" for a particular sound, you are effectively turning it off.

Problem: I'm told that people can hear my voice and other noises in the room when I am transmitting.

Solution: Unplug your microphone. You're feeding your PSK31 signal to the accessory jack on your transceiver and probably keying it that way as well. The problem is that some of your radio is still accepting audio from your microphone jack, too. When you key the radio to transmit PSK31, your microphone is live!

Problem: Stations report that my PSK31 signal looks fuzzy on their waterfall displays.

Solution: When your PSK31 station is idling (not sending data), stations should see your signal as two distinct lines, possibly with a bit of fuzziness between. When you start typing, the lines should look as though they are segmented or twisted together. A PSK31 signal that looks like a solid fuzzy line often indicates the presence of distortion. RF may be getting into your sound card, the sound card may be overdriven, or you could be picking up hum induced by ground loops. Try reducing output and see if that cures the problem. If not, consider using a couple of isolation transformers in the cables between the sound card and your radio.

Problem: The PSK31 software doesn't key my radio into the transmit mode.

Solution: Most hams are using some variation of the circuit shown in Figure 2-3 (in Chapter 2) to key their radios from their computer COM ports. If this applies to you, don't forget that some PSK31 programs activate the COM port DTR or RTS pins—but not both. (The original G3PLX software sends logic "highs" to both pins.) If you have your keying interface tied to the DTR (or RTS) pin, try switching it to the other pin.—*WB8IMY*

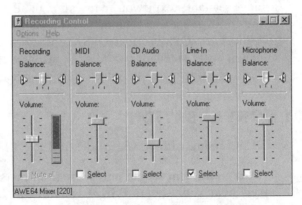

This is a typical *Windows* audio mixer panel for the recording controls. Depending on the software or soundcard you are using, your mixer may have a different appearance. The recording mixer controls the audio signals going *to* the soundcard. In this example, you can see that the "Line-in Balance" control is selected and the "slider" is at the top. If I unchecked the "Select" box, or brought the slider down to the bottom, the soundcard would *not* process the received audio from the transceiver. If your PSK31 software doesn't seem to receive, this is one of the first things to check.

PSK31 signals have a distinctive sound unlike any digital mode you've heard on the ham bands. You won't find PSK31 by listening for the *deedle-deedle* of a RTTY signal, and PSK31 doesn't "chirp" like the TOR modes. PSK31 signals *warble*—that's the best way I can describe them.

One remarkable aspect of *DigiPan* is that it allows you to see (and often copy) PSK31 signals that you cannot otherwise hear. It is not at all uncommon to see several strong signals (the audible ones) interspersed with wispy blue ghosts of very weak "silent" signals. I've clicked on a few of these ghosts and have been rewarded with text (not error-free, but good enough to understand what is being discussed).

Using *DigiPan* reminds me of the sonar operators in the movie *The Hunt for Red October*. There is an eerie excitement in finding one of those ghostly traces and muttering to yourself, "Hmmm...what do we have here? An enemy submarine rigged for silent running? A distant pod of killer whales? Or Charlie in Sacramento running 5 W to his attic dipole?"

Spend some time tracking down PSK31 signals and watching the conversations. With a little practice you'll discover that tuning becomes much easier. You'll also find that you develop an "ear" for the distinctive PSK31 signal.

You may occasionally stumble across a conversation taking place with QPSK. The tones sound the same, but you'll find it impossible to achieve readable text if you're tuning in the BPSK mode. Try switching to the QPSK mode and retuning. And remember that sideband selection is critical for QPSK. Most QPSK operators choose upper sideband, although lower sideband would work just as well. The point is that *both* stations must be using the same sideband, whether it's upper or lower.

A PSK31 CONVERSATION

Conversing with PSK31 is identical to RTTY. For example:

Yes, John, I'm seeing perfect text on my screen, but I can barely hear your signal. PSK31 is amazing! KF6I DE WB8IMY K

I know what you mean, Steve. You are also weak on my end, but 100% copy. WB8IMY DE KF6I K

Some PSK31 programs and processor software offer type-ahead buffers, which allow you to compose your response "off line" while you are reading the incoming text from the other station. The original PSK31 software for *Windows* lacked this feature, although it did allow you to send "canned" pre-typed text blocks known as *brag files*. (Brag files are usually descriptions of your station setup.)

PSK31 conversations flow casually, just like RTTY. The primary difference is that you will usually experience perfect or near-perfect copy under conditions that would probably render RTTY useless.

IS PSK31 THE HEIR TO THE HF DIGITAL THRONE?

The best answer is a strong "maybe." I'm not brave enough to stick my neck out and predict that PSK31 will overtake RTTY as the number one "live" HF digital mode, but it is off to a promising start. Even though Amateur Radio is a technological hobby, hams embrace change reluctantly. Will the graybeards of RTTY forsake their *deedle-deedles* for PSK31? Stay tuned.

Chapter FIVE

PACTOR

Hans-Peter Helfert, DL6MAA, and Ulrich Strate, DF4KV, developed PACTOR in 1991. It remains to this day the most popular of the HF *burst* modes. We call PACTOR a burst mode because of the way it sends information. Rather than sending a continuous stream of data like RTTY, PSK31, MFSK16 or Hellschreiber, PACTOR transmits bursts of information that take the form of short data blocks. When the data is received intact, the receiving station sends an *ACK* signal (for *acknowledgment*). If the data contains errors, a *NAK* is sent (for *nonacknowledgment*). In simple terms, ACK means, "I've received the last group of characters okay. Send the next group." NAK means, "There are errors in the last group of characters, send them again." This back-and-forth data conversation sounds like crickets chirping. In the case of PACTOR, the long chirp is the data and the short chirp is the ACK or NAK.

MEMORY ARQ

In less sophisticated burst modes such as AMTOR or packet, a data block must be repeated over and over if that's what it takes to deliver the information error-free. This results in slow communication, especially when conditions are poor.

PACTOR handles the challenge of "repairing" errors in an interesting way. Each data block is sent and acknowledged with an ACK signal if it's received intact. If signal fading or interference destroys some of the data, a NAK is sent and the block is repeated. Nothing new so far—packet and

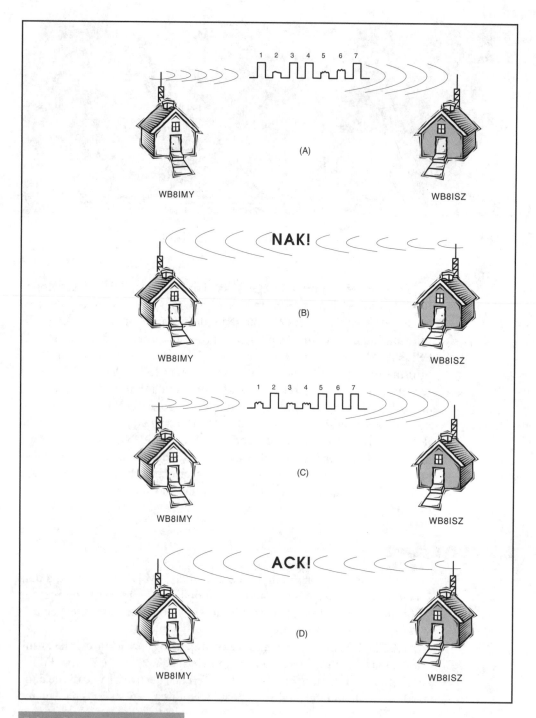

(A)

WB8IMY

WB8ISZ

NAK!

(B)

WB8IMY

WB8ISZ

(C)

WB8IMY

WB8ISZ

ACK!

(D)

WB8IMY

WB8ISZ

AMTOR behave much the same way. The big difference, however, involves *memory*.

When a PACTOR controller receives a mangled character block, it analyzes the parts and temporarily memorizes whatever information appears to be error-free. If the block is shot full of holes on the next transmission as well, the controller quickly compares the new data fragments with what it has memorized. It fills the gaps as much as possible and then, if necessary, asks for another repeat. Eventually, the controller gathers enough fragments to construct the entire block (see Fig 5-1). PACTOR's *memory ARQ* feature dramatically reduces the need to make repeat transmissions of damaged data. This translates into much higher throughput.

PACTOR has the capability to communicate at varying speeds according to band conditions. Under good conditions, PACTOR will accelerate to 200 baud. PACTOR throughput is enhanced by using *Huffman* coding that reduces the average character length for improved efficiency. Table 5-1 (at the end of this chapter) shows the Huffman compressed equivalent of each ASCII character used in PACTOR, with the least significant bit (LSB) given first. The length of individual characters varies from 2 to 15 bits, with the most frequently used characters being the shortest. This results in an average character length of 4 to 5 bits for English text, instead of the 8 bits required for normal ASCII.

Fig 5-1—Memory ARQ at work. WB8IMY sends data to WB8ISZ (A), but bits 2, 5 and 6 are corrupted. The controller at WB8ISZ's station memorizes the good data as well as the positions of the corrupted bits. It sends a NAK (B) to demand a repeat of all the data. On the next transmission (C), bits 1, 3 and 4 are corrupted, but that's not a problem. WB8ISZ's controller has these bits stored in memory from the first transmission. What's important is the fact that bits 2, 5 and 6 made it through intact. They're the *missing bits* from the first transmission. These bits are combined with the ones in memory and the entire data segment is complete! WB8ISZ now sends an ACK (D) for the next batch. (A PACTOR data segment is actually 192 bits long when operating at 200 bit/s; 80 bits long at 100 bit/s.)

WHAT DO I NEED TO RUN PACTOR?

Assembling a PACTOR station is very simple. All you need are the following:

• An SSB transceiver. PACTOR uses a two-tone mark/space system similar to RTTY. This means that PACTOR can be sent with AFSK, with transmit audio applied at the microphone or accessory jack, or with FSK, where the transmitter generates the mark/space frequency shifts by itself. If you operate in AFSK, it doesn't matter which sideband you use, upper or lower.

• A data terminal or a computer running terminal software

• A multimode communications processor (MCP) with PACTOR capability. Every MCP manufactured today includes PACTOR.

When assembling your PACTOR station, use the guidelines discussed in Chapter 2. If you're already set up with an MCP for the other digital modes, you don't need extra equipment to operate PACTOR.

You'll notice that I didn't mention using a soundcard modem for PACTOR. When this book was written, Brian Beezley, K6STI, had developed the only soundcard-based program that included PACTOR—*RITTY*. That software has been discontinued, however, and no one has yet

Dissecting a PACTOR Data Packet

A PACTOR data packet is either 96 bits long sent at 100 baud, or 192 bits long sent at 200 baud, with the data rate dependent on conditions. Each packet consists of a Header byte, Data field, and Status byte, followed by the CRC byte given twice. The Header byte consists of the 8-bit pattern for 55 hexadecimal and is used for synchronization, Memory-ARQ, and listen mode. The Data field contains 64 bits if sent at 100 baud, or 160 bits if sent at 200 baud. Its format is normally Huffman-compressed ASCII, with conventional 8-bit ASCII as the alternative. The Status byte provides the packet count, data format (whether standard 8-bit ASCII or Huffman-compressed ASCII), break-in request, and QRT bit, for a total of 8 bits. The CRC calculation is based on the ITU-T polynomial xE16+xE12+xE5+1. The CRC byte is calculated for the whole packet starting with the data field, without Header, and consists of 16 bits.

stepped forward to write a new PACTOR program for PC soundcards. No doubt that situation will change in the near future.

CQ PACTOR

You call CQ on PACTOR using *forward error correction*, or *FEC*. The FEC signal sounds like very fast RTTY, but in reality it is a stream of data in which each character is repeated twice for redundancy—there are no ACKs or NAKs. Obviously, this means that an FEC transmission is not error-free, but the copy is good enough to pull out the call sign of the sending station. When you're sending a PACTOR CQ, that's really all that matters.

If you hear a signal that you suspect is a PACTOR FEC CQ, tune carefully until your MCP indicates that it has locked (synchronized) with the FEC signal. Within a short time you should begin to see text on your screen.

CQ CQ CQ DE WB8IMY WB8IMY WB8IMY
CQ CQ CQ DE WB8IMY WB8IMY WB8IMY
CQ CQ CQ DE WB8IMY WB8IMY WB8IMY K K

Notice that this CQ uses several short lines of text rather than a few long lines. This helps stations synchronize more easily.

Starting a Conversation

Answering a CQ in PACTOR is straightforward. Depending on the software you're using, it may be as simple as entering:

CALL N1BKE
or, **CONNECT N1BKE**

…at the **cmd:** prompt. Some types of software streamline the process even further. There may be pop-up boxes where you simply enter a call sign.

Once you make contact, the conversation proceeds in turns. This means that one station "talks" (the *ISS* or *information sending station*) while the other "listens" (the *IRS* or *information receiving station*). When the ISS has had his say, he sends a special control signal known as the *over* command that immediately reverses the roles—suddenly you are the ISS and he is the IRS. Introduce yourself and ask a question about where he lives, or what he does for a living. Use the over command to flip the link again. A conversation is underway!

Depending on the software you are using with your MCP, sending the over command is usually as easy as tapping a single key. The software usually provides some sort of visual indicator to show which mode you are in—IRS or ISS—in case you become confused! If you forget to send the *over* command, your stations will simply sit there and chirp mindlessly at each other. Fortunately, the IRS can send an over command and flip the link for you if necessary. This is known as a "forced over."

THE EVOLUTION OF THE PACTOR BBS

In the early days of PACTOR, live conversations were common. This is no longer the case because amateurs prefer to use RTTY or PSK31 for

PACTOR II Frames

The basic PACTOR-II frame structure is similar to PACTOR-I. It consists of a header, a data field, the status byte and the CRC. The standard cycle duration is 1.25 seconds—the same as PACTOR. Even so, the lengths of the control signals, such as the ACKs and NAKs, were increased somewhat to improve reliability under poor conditions. As a result, the length of the data packet had to be shortened to 800 ms so that everything would still "fit" within the 1.25-second cycle. PACTOR II also has a "long path" mode similar to PACTOR where the cycle is stretched to accommodate signal returns from very distant stations.

Unlike PACTOR, PACTOR II can switch to longer packets if the data blocks are not filled up with idles (i.e., if the transmitter buffer indicates that more information has to be transferred than fitting into the standard packets). If the information sending station (ISS) prefers to use long packets, it automatically sets the long-cycle flag in the status word. The information receiving station (IRS) then can accept the change and away you go! Now the data field can contain as many as 2208 bits of information. These data packets are 3.28 seconds in length—quite large by HF digital standards—and the entire cycle duration switches to 3.75 seconds. This mode is primarily used when you are exchanging large files with another station.

live keyboard-to-keyboard chats. PACTOR instead has evolved into a favored mode for contacting automated mailboxes and bulletin board systems (BBSs).

Many PACTOR BBSs in the early '90s acted as HF e-mail gateways to the VHF amateur packet radio network. You could connect to a PACTOR BBS on, say, 20 meters, and post a message that would eventually reach a distant packet BBS on 2 meters.

With the advent of affordable Internet access, amateur VHF packet

PACTOR II Data Compression

PACTOR and PACTOR II both use automatic on-line Huffman data compression. PACTOR II, however, also uses run-length encoding and Pseudo-Markov Compression (PMC). Compared to 8-bit ASCII (plain text) PMC yields a compression factor of around 1.9, which leads to an effective speed of about 600 bits per second in average propagation conditions in data mode. The PACTOR II processor automatically evaluates the condition of the link from moment to moment and selects PMC, Huffman encoding or the original ASCII code, whichever is best at the time.

Ordinary Huffman compression exploits the *one-dimensional* probability distribution of the characters in plain texts. The more frequently a character occurs, the shorter the Huffman symbol assigned to it. Markov compression, on the other hand, can be considered as a "double" Huffman compression because for each preceding character, a probability distribution of the very next character can be calculated. For example, if the actual character is "e," it is very likely that "i" or "s" occurs next, but extremely unlikely that a "w" follows.

As you can probably guess, Markov coding would require an enormously large and complex lookup table—well beyond the capacity of a stand-alone communications processor! That is why SCS chose Pseudo Markov Coding. With PMC the Markov encoding is limited to the 16 most frequent preceding characters. All other characters trigger normal Huffman compression of the very next character. This reduces the Markov coding table to a reasonable size and also makes the character probabilities less critical.

networks declined substantially to the point where they could no long handle messages reliably. To adapt to the new rules of the game, a number of PACTOR BBSs evolved into Internet e-mail gateways. These HF e-mail outlets have proven popular among amateurs who find themselves in distant locations without Internet access: sailing enthusiasts, missionaries, adventurers, etc. By connecting to a PACTOR BBS on HF, they can send and receive Internet e-mail with ease. HF Internet gateways can also be found running Clover and PACTOR II. See Chapter 9 for a more detailed discussion.

Not all PACTOR BBSs are gateways. Some are simply clearinghouses for information. It's fun to check into a BBS and explore what it has to offer.

When you connect to a BBS you usually see a command line similar to the one shown below.

Welcome to WB8SVN's PACTOR BBS in Placentia, CA. Enter command: A,B,C,D,G,H,I,J,K,L,M,N,P,R,S,T,U,V,W,X,?,* >

After you enter a command you *do not* need to send the over code. The BBS will flip the link automatically after it receives the command from you.

You can use PACTOR BBSs to exchange messages with other PACTOR operators. And with PACTOR's ability to handle binary data, you can even download small programs from the BBS and run them on your computer!

PACTOR II

The state of the art never remains fixed. This is as true of PACTOR as it is of any communication mode. At the 1993 Dayton HamVention the inventors of PACTOR—now incorporated as Special Communications Systems (SCS) of Hanau, Germany—unveiled their newest creation: *PACTOR II*.

You can think of PACTOR II in terms of being a supercharged version of PACTOR (or perhaps we should refer to it as PACTOR I!). In good conditions PACTOR II boasts a data transfer rate up to six times faster than PACTOR. At the same time, PACTOR II is capable of maintaining links in conditions where the signal-to-noise ratio is at −18 dB. This means that PACTOR II can carry on an exchange even when the signals are virtually inaudible. Such remarkable performance is also conservative when it comes to bandwidth; a PACTOR II signal only occupies about 500 Hz of spectrum.

The complex *pi/4-DQPSK* modulation system used by PACTOR II requires DSP technology and fast micro-processors. With DSP you let the *software* decide the composition of the signal, not the hardware. DSP is much more flexible, allowing you to create the signal you want without having to worry about hardware filters and so on. The tradeoff is that you must use high-speed microprocessors to handle all the incoming and outgoing information at a decent rate.

PACTOR II is also *backward compatible* with PACTOR. That is, a PACTOR II operator can communicate with a PACTOR operator and vice versa. Many HF digital BBSs, including Internet e-mail gateways, include PACTOR II among their operating modes.

PACTOR II is only available in the PTC-II series multimode processors manufactured (or licensed) by SCS. In addition to PACTOR II, the PTC-II offers other HF modes such as RTTY, SSTV, packet and even PSK31.

The primary markets for PACTOR II have been commercial and military, and the PTC-II was designed and priced accordingly. At a cost of nearly $1000 each, hams didn't exactly flock to PTC-II processors in great numbers. In 1999, however, SCS introduced the PTC-IIe processor (see Chapter 1). Selling at less than $400, this multimode box may finally attract a larger amateur audience for PACTOR II. Of course, it has stiff price and performance competition from Clover, which we'll discuss in the next chapter.

PACTOR II STATION REQUIREMENTS AND OPERATION

The requirements for a PACTOR II station are essentially the same as for a PACTOR station: An SSB transceiver, a computer or data terminal, and a PTC-II or PTC-IIe processor.

PACTOR II also operates in nearly the same fashion as PACTOR. In fact, the link is initially established using the PACTOR protocol, then automatically switched to PACTOR II if both stations are using PACTOR II processors. And as with PACTOR, you must take turns during the conversation, sending an *over* command to allow the other operator to send data.

QST REVIEWS THE SCS PTC-IIE PACTOR II PROCESSOR

(from the April 2000 issue of QST*)*

The SCS PTC-IIe is the smaller sized, and smaller priced, brother of the PTC-II controller. The PTC and PTC-IIe are the only MCPs that offer PACTOR II—and the PTC-IIe is the only MCP to date that includes PSK31 in its list of modes. SCS invented both PACTOR and PACTOR II, the most widely used HF "burst" modes in Amateur Radio. PACTOR was a hit almost from the moment it appeared in the early '90s, but PACTOR II has been slower to catch on. This has been primarily due to the $800-900 price tag of the PTC-II controller. The PTC-IIe offers greater economy, but at $649 it is still almost equal to the cost of a low-end HF transceiver. Is it worth it?

The PTC-IIe's PACTOR II performance was astonishing. In several instances we successfully accessed BBSs under abominable conditions where we could scarcely hear the other station's signals in the noise. With PACTOR II the data flowed smoothly, albeit more slowly in the midst of deep fades. On the basis of on-air tests, the PTC-IIe implementation of PACTOR II clearly outperformed G-TOR. When we put it up against Clover II, however, the differences were less obvious. In fact, it was impossible to determine a "winner" between the two modes.

In PSK31 the PTC-IIe was mediocre. We matched it head-to-head against sound card-based PSK31 software and the sound cards came out on top every time. Part of the problem may have involved the level of tuning precision required for PSK31. The PTC-IIe uses its multisegment LED bargraph in an interesting way, adjusting the intensity of the segments until the center three are glowing brightest when the PSK31 signal is properly tuned. Clever as this may be, it seemed as though we were able to achieve more accurate tuning, and better copy, using the more common PSK31 software "waterfall" displays. RTTY performance was very good, better than either the KAM '98 or PK232DSP, but not quite as sharp as the HAL DXP38. One of our reviewers tried the PTC-IIe during the 2000 ARRL RTTY roundup and was impressed with its ability to copy weak signals in strong interference. The PTC-IIe works strictly in AFSK; it does not provide an FSK output.

The PTC-IIe offers both HF and VHF packet, but the manual is vague about this feature. It spends several pages discussing 300-baud (HF) packet—correctly disdaining it as a poor mode for HF work—with hardly a mention of 1200-baud (VHF) packet. With a little experimentation we had the PTC-IIe working nicely at 1200 baud. The only drawback is that the unit is not GPS compatible for use with the Automatic Position Reporting System.

The DSP filtering function of the PTC-IIe can be put to work for other applications. By entering the *Audio* mode you can use the PTC-IIe in the same way you would an external DSP audio filter. We fed the audio to an amplifier and listened as we experimented with the notch filter, noise filter, etc. All adjustments are made from the keyboard. This is a nifty feature for those times when you'd like to take a break from digital and work a little phone or CW.

Speaking of CW, yes, the PTC-IIe does that, too. Once again, the copy ability was similar to the other MCPs we tested. The PTC-IIe also does SSTV and fax, but for these modes it functions more like a modem and requires external software to do the actual signal processing and display. Installation of the PTC-IIe is fairly simple. On the rear panel there is a DIN jack for the audio in/out and transceiver keying lines. Connect the lines as shown in the manual and you are in business. Audio output levels are controlled through software commands.

The English manual has been translated well from the original German, but the organization is poor. It is not easy to locate the information you need, and you often find it lacking in detail. The software provided with the PTC-IIe is *DOS* only, and it performed reasonably well. At this price level, however, we had expected to find a full-featured *Windows* program.

So the question arises once again—is the PTC-IIe worth it? The PTC-IIe is targeted to a particular market: the traveling hams. These are the folks who cruise the roads for months in recreational vehicles or roam the seas in private boats. Many of these travelers depend on the global Winlink 2000 BBS network to stay in touch via Internet e-mail. The great majority of these Winlink gateway stations using PACTOR or PACTOR II (usually both). So, it is important to have a high-performance PACTOR/PACTOR II MCP that can maintain connections under the worst conditions. When e-mail is one of your lifelines to friends and loved ones, price is less of a consideration than quality. For the average housebound ham, $649 may be too steep to justify—even for the remarkable performance of PACTOR II. For the traveling amateur, however, that price tag is quite reasonable.

Table 5-1
PACTOR Huffman Compression

Char	ASCII	Huffman (LSB [sent first] on left)
space	32	10
e	101	011
n	10	0101
i	105	1101
r	114	1110
t	116	00000
s	115	00100
d	100	00111
a	97	01000
u	117	11111
l	108	000010
h	104	000100
g	103	000111
m	109	001011
<CR>	13	001100
<LF>	10	001101
o	111	010010
c	99	010011
b	98	0000110
f	102	0000111
w	119	0001100
D	68	0001101
k	107	0010101
z	122	1100010
.	46	1100100
,	44	1100101
S	83	1111011
A	65	00101001
E	69	11000000
P	112	11000010
v	118	11000011
O	48	11000111
F	70	11001100
B	66	11001111
C	67	11110001
I	73	11110010
T	84	11110100
O	79	000101000
P	80	000101100
1	49	001010000
R	82	110000010
(	40	110011011
)	41	110011100
L	76	110011101

Char	ASCII	Huffman (LSB [sent first] on left)
N	78	111100000
Z	90	111100110
M	77	111101010
9	57	0001010010
W	87	0001010100
5	53	0001010101
y	121	0001010110
2	50	0001011010
3	51	0001011011
4	52	0001011100
6	54	0001011101
7	55	0001011110
8	56	0001011111
H	72	0010100010
J	74	1100000110
U	85	1100000111
V	86	1100011000
<FS>	28	1100011001
x	120	1100011010
K	75	1100110100
?	63	1100110101
=	61	1111000010
q	113	1111010110
Q	81	1111010111
j	106	00010100110
G	71	00010100111
-	45	00010101111
:	58	00101000111
!	33	11110011101
/	47	11110011110
*	42	001010001100
"	34	110001101100
%	37	110001101101
'	39	110001101110
_	95	111100001100
&	38	111100111001
+	43	111100111110
>	62	111100111111
@	64	0001010111000
$	36	0001010111001
<	60	0001010111010
X	88	0001010111011
#	35	0010100011011
Y	89	00101000110101
;	59	11110000110100
	92	11110000110101
[	91	001010001101000
]	93	001010001101000

Char	ASCII	Huffman (LSB [sent first] on left)
<DEL>	127	110001101111000
~	126	110001101111001
}	125	110001101111010
	124	110001101111011
{	123	110001101111100
'	96	110001101111101
^	94	110001101111110
<US>	31	110001101111111
<GS>	29	111100001101100
<ESC>	27	111100001101101
<EM>	25	111100001101110
<CAN>	24	111100001101111
<ETB>	23	111100001110000
<SYN>	22	111100001110001
<NAK>	21	111100001110010
<DC4>	20	111100001110011
<DC3>	19	111100001110100
<DC2>	18	111100001110101
<DC1>	17	111100001110110
<DLE>	16	111100001110111
<RS>	30	111100001111000
<SI>	15	111100001111001
<SO>	14	111100001111010
<FF>	12	111100001111011
<VT>	11	111100001111100
<HT>	9	111100001111101
<BS>	8	111100001111110
<BEL>	7	111100001111111
<ACK>	6	111100111000000
<ENQ>	5	111100111000001
<EOT>	4	111100111000010
<ETX>	3	111100111000011
<STX>	2	111100111000100
<SOH>	1	111100111000101
<NUL>	0	111100111000110
<SUB>	26	111100111000111

Chapter SIX

CLOVER

According to legend, the late Ray Petit, W7GHM, was working on a revolutionary new HF digital communication system in his home lab. His wife stopped by just in time to see a display of the received signal pattern on Ray's oscilloscope. "That looks just like a Clover," she remarked. The new mode didn't have a name and his wife's description seemed appropriate enough. Why not simply call it *Clover*?

WHAT IS CLOVER?

Clover is an advanced HF digital communication system which Ray developed in a joint venture with HAL Communications of Urbana, Illinois. Clover uses a four-tone modulation scheme. An enhanced version using improved DSP technology came along later and was officially christened Clover II, but I'll refer to the mode simply as "Clover" throughout this chapter for the sake of clarity.

Depending on signal conditions, several different modulation formats can be selected manually or automatically. Each tone is phase- and/or amplitude-modulated as a separate, narrow-bandwidth data channel. As you might guess, the resulting Clover signal is very complex!

For example, when the tone pulses are modulated using quadrature phase-shift modulation (QPSM), the differential phase of each tone shifts in 90° increments. Two bits of data are carried by each tone for a total of eight bits in each 32-ms frame. The resulting block data rate is about 250 bits per second. Clover is capable of even higher data rates when using 16-phase, four-amplitude modulation (16P4A). In this format, Clover perks along at 750 bit/s.

The complex, higher-speed modulation systems are used when conditions are favorable. When the going gets rough, Clover automatically brings several slower (but more robust) modes into play.

Even with these ingenious adaptive modulation systems, errors are bound to occur. That's where Clover's Reed-Solomon coding fills the gaps. Reed-Solomon coding is used in all Clover modes. Errors are detected at the receiving station by comparing check bytes that are inserted in each block of transmitted text. When operating in the ARQ (automatic repeat request) mode, Clover's damaged data can often be reconstructed without the need to request repeat transmissions (sort of a nonARQ type of ARQ!). This is a major departure from the techniques used by PACTOR. Of course, Clover can't always repair data; repeat transmissions—which Clover handles automatically—are sometimes required to get everything right.

With the combination of adaptive modulation systems and Reed-Solomon coding, Clover boasts remarkable performance—even under the worst HF conditions. The only Amateur Radio digital mode with the

The Clover Waveform

The Clover waveform consists of four tone pulses, each of which is 125-Hz wide, spaced at 125-Hz centers. The four tone pulses are sequential, with only one tone being present at any instant and each tone lasting 8 ms. Each frame consists of four tone pulses lasting a total of 32 ms, so the base modulation rate of a Clover signal is always 31.25 symbols per second. Data is conveyed by changing the phase and/or amplitude of successive pulses at the same frequency. These changes are made only at the instant midway between the peaks of two successive pulses when their amplitudes are zero. The measured Clover modulation spectra is tightly confined within a 500 Hz bandwidth, with outside edges sup-pressed 50 dB to prevent interference to adjacent frequencies. Unlike other modulation schemes, the Clover modulation spectra is the same for all modulation formats. Additional key parameters of Clover modulation include a symbol rate of 31.25 symbol/s (regardless of the type of modulation being used), 2:1 voltage (6 dB power) crest factor, and an ITU-R emission designator of 500H J2 DEN or 500H J2 BEN.

potential to match Clover's performance is PACTOR II, and possibly G-TOR (under certain conditions).

CLOVER HANDSHAKING

As you may recall, PACTOR uses an *over* command to switch the link so that one station can send while the other receives. Clover links must be switched as well, but the switching takes place *without* using *over* commands.

When two Clover stations make contact, they can send limited amounts of data to each other (up to 30 characters in each block) in what is known as the *chat mode*. If the amount of data waiting for transmission at one station exceeds 30 characters, Clover automatically switches to the *block data mode*. The transmitted blocks immediately become larger and are sent much faster. The other station, however, remains in the chat mode. Because of precise frame timing, all of this takes place without the need for either operator to change settings, or send *over* commands. The Clover controllers at both stations "know" when to switch from transmit to receive and vice versa. And what if both stations have large amounts of data to send at the same time? Then they *both* switch to the data block mode. This high degree of efficiency is transparent to you, the operator.

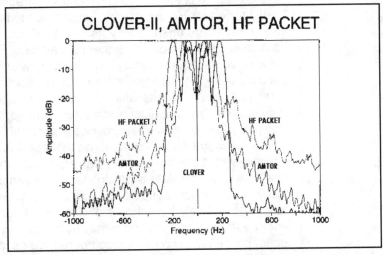

The four-tone Clover waveform compared to HF packet and AMTOR waveforms.

All you have to do is type your comments or select the file you want to send—Clover takes care of everything else!

Clover features an FEC mode similar to that used by PACTOR and G-TOR. You use the Clover FEC to call CQ, or to send transmissions that can be received by several stations at once. (In the Clover ARQ mode, only two stations can communicate at a time.)

WHAT DO I NEED TO RUN CLOVER?

The requirements for a Clover station differ somewhat from those of other HF digital modes. They are:

• An SSB transceiver. The transceiver must be stable (less than 30 Hz drift per hour). It should also include a frequency display with 10-Hz

---

The Reed-Solomon Error Detection and Correction Algorithm

Reed-Solomon FEC is used in all Clover modes. This is a powerful byte and block oriented error-correction technique, not available in other common HF data modes, and it can allow the receiving station to correct errors without requiring a repeat transmission. Errors are detected on octets of data rather than on the individual bits themselves. This error correction technique is ideally suited for HF use in which errors due to fades or interferences are often "bursty" (short-lived) but cause total destruction of a number of sequential data bits. Error correction at the receiver is determined by "check" bytes which are inserted in each block by the transmitter. The receiver uses these check bytes to reconstruct data which has been damaged during transmission. The capacity of the error corrector to fix errors is limited and set by how many check bytes are sent per block. Check bytes are also "overhead" on the signal and their addition effectively reduces the efficiency and therefore the "throughput rate" at which user data is passed between transmitter and receiver. Efficiencies of 60%, 75% or 90% can be invoked by using successively lower levels of Reed-Solomon encoding for error correction, or 100% efficiency by bypassing this algorithm. Better propagation conditions do not require as much error correction, which means the amount of overhead decreases and the efficiency increases.

How CLOVER Works

By Ray Petit, W7GHM (SK)

To adequately explain why CLOVER is such a breakthrough, we must first briefly review the pluses and minuses of existing HF data modes—RTTY, AMTOR, and HF packet radio.

RTTY of course led the way for automatic reception of characters or data via HF radio. RTTY has been around since the 1940s and is very reliable. The techniques we use today to send and receive RTTY are much the same as those first used. We have better equipment, but use the same FSK modulation and Baudot or ASCII code. RTTY is slow and does not offer any error correction or detection. RTTY speeds of 60 WPM (45 baud) to 100 WPM (75 baud) are common. Increasing the RTTY spreads increases the errors; we generally stick to 45 baud.

AMTOR evolved from an existing ship-to-shore "radio telex" mode, often called "TOR" or "SITOR" (CCIR 4767 and CCIR 625). AMTOR introduced us to a new type of data link — "ARQ mode" (ARQ stands for Automatic Repeat Request).

AMTOR characters are coded so that the receiving station can detect an error in each character sent. The sending station sends three characters, turns the transmitter OFF, and listens for a character response from the receiving station. The response is either "all OK, send next three," or "repeat last three characters." Like RTTY, it is also pretty "slow." Under the best conditions, AMTOR can pass data at an equivalent RTTY rate of 50 baud (6.67 characters per second). AMTOR is also limited to the same character set as Baudot — just all capital letters and no ASCII control characters.

CLOVER intends to support the many advantages of AMTOR and HF packet radio and fixes the major problems of these modes. The most serious limitation of RTTY, AMTOR, and HF packet is data throughput and how the data is used to modulate the radio signal. The ionosphere is not a friendly medium for data signals. HF signals often arrive at the receiving antenna by many different propagation paths; two or more paths are common. Each signal path has its own time delay, amplitude, and even different center frequency. The receiving antenna does not discriminate; it adds all the signals and passes the composite to the receiver. The amplitudes and phases of the separate ac signals combine algebraically to produce a widely varying receiver input. Deep selective fades and time smearing of the data pulse transitions are the usual result.

Once combined at the antenna, the individual path signals are not easily separated. It is usually impossible to compensate for all of these "multipath" effects in the demodulator. A good example of multipath ionosphere distortion is the "selective fading" we hear when listening to music from a shortwave radio station. While annoying when listening to music, the distortion can be totally destructive to data transmissions.

A major nonrecoverable parameter of HF data is the time at which the data state changes from MARK to SPACE, the data transition time. If we lose this information, the modem can no longer tell when one data pulse ends and the next one begins or if the logic state should be a "1" or a "0." When two signals arrive with different propagation time delays, the composite antenna output signal is "smeared" and the transition times overlap. Measurements by Ray and many others show that we can expect this time overlap from different paths to be as much as 3 to 5 milliseconds (ms). At least one half of each data pulse without distortion determines the MARK or SPACE data state. Therefore, the narrowest data pulse which can be reliably demodulated is on the order of 6 to 10 ms, corresponding to maximum data rates in the range of 100 to 167 baud. Observation shows that the 100-baud limit is more realistic and even it can be too high for satisfactory data transmission at times.

HF packet radio uses a 300-baud data rate, a pulse width of 3.3 milliseconds. Successful HF packet transmissions are therefore very unlikely if the signal is propagated by multiple paths. HF packet works well only when the operating frequency is close to the Maximum Usable Frequency (MUF)—when there is only one propagation path. Since this is the exception and not the rule, long-term packet performance on a single fixed frequency is pretty poor, and many repeats may be required to pass any data at all.

HF packet radio, AMTOR, and RTTY all use FSK modulation. One radio frequency is sent for the "1" or MARK pulse state and another for the "0" or SPACE state. The transmitter carrier frequency is shifted back and forth at the same rate as the data. CLOVER uses different modulation techniques. First, CLOVER shifts the phase and not the frequency of the carrier. Second, more than one bit of data can be sent per phase state. For example, BPSK (binary phase shift keying) has two phase states (0 or 180 degrees) which can be used to represent MARK and SPACE. QPSK (Quadrature PSK) has four phase states (0, 90, 180, 270 degrees). A single-phase change in QPSK represents the state of two binary bits of data. Similarly, 8PSK can send the state of 3 bits per phase change and 16PSK can send 4 bits per phase change.

CLOVER also allows use of Amplitude Shift Keying (ASK) in the 8PSK and 16PSK modes. We call these modes "8P2A" (4 data bits per phase/amplitude change) and "16P4A" (6 bits per phase/amplitude change.) Since all changes in phase or amplitude occur at the fixed base rate of 31.25 BPS (an equivalent pulse width of 32 ms), data errors due to multipath time smearing of data transitions are minimized.

The CLOVER modulation strategy is to always send data at a very slow base modulation rate and to use multi-level changes in phase or amplitude to speed up data flow. One final twist to CLOVER-II is that there are four separate transmitted pulses, each separated by 125 Hz.

Each of the pulses may be modulated by BPSK through 16PSK plus 8P2A or 16P4A modulation. This further multiplies that effective throughput by a factor of four. Putting it all together, CLOVER can send data at rates from its base data rate (31.25 bps) to 24 times its base rate (750 bps). Wow! It's almost like getting something for nothing! Not quite so. There are still problems to be solved.

PSK modulation itself poses some pretty serious problems. IF we modulate a continuous carrier using PSK, the frequency spectrum we get is very bad for HF use, as sidebands are strong and extend over a wide spectrum. CLOVER avoids this problem by two techniques:

(1) Each of the four tones is an ON/OFF amplitude pulse and the phase is changed only when the pulse is OFF;

(2) The amplitude waveform of each ON/OFF pulse is carefully shaped to minimize the resulting frequency spectrum.

Combined, these techniques produce a composite CLOVER spectrum that is only 500 Hz wide down to −60 dB. This is one half the radio bandwidth required for AMTOR and one quarter that for HF packet radio at an effective data rate of up to 100 times faster than HF packet radio or AMTOR!

CLOVER also takes a different approach to error correction. AMTOR and packet radio both correct errors by sensing errors at the receiver and then requesting repeat transmissions. When there are errors to be fixed, data throughput is slowed by the time it takes to send the repeats. When conditions are getting poor, packet radio bogs down to sending only repeats and no data at all; AMTOR will slow down considerably under the same conditions.

CLOVER uses a Reed-Solomon error correction code that allows the receiver to actually fix errors WITHOUT requiring repeat transmissions.

For a moderate number of errors, CLOVER doesn't require repeats and the data continuous flowing at the no-error rate. To distinguish between the two schemes, we classify AMTOR and packet radio as "error-detection" protocols and CLOVER as an "error-correction" protocol. In addition, like packet radio, CLOVER includes a CRC (Cyclic Redundancy Check sum) which is used when conditions are very bad and the number of errors exceeds the capacity of the Reed-Solomon error corrector.

CLOVER ARQ mode is also adaptive. As a result of the DSP calculations necessary to detect multi-level PSK and ASK, the CLOVER receiver already has information which can be used to determine the signal-to-noise ratio (S/N), phase dispersion, and time dispersion of the received signal. CLOVER has 8 different modulation modes, 4 different error correction settings, and 4 different data block lengths that can be used—a total of 128 different modulation/code/block combinations!

Using real-time signal analysis, the CLOVER receiver will automatically signal the transmitting state to change modes to match existing ionosphere conditions. When propagation is very good, CLOVER can set itself to the higher speed and data literally "screams" down the path.

When conditions are not so great, the data speed is slowed. As noted earlier, the CLOVER character throughput rate under typical HF conditions is about ten times faster than AMTOR or HF packet, However when we get one of those perfect conditions, CLOVER will "shift gears" and pass data at 50 to 100 times the speed of AMTOR of HF packet radio. In all cases, CLOVER automatically changes speeds to give the maximum speed that the ionosphere will allow.

resolution. The audio output from the Clover controller is fed to the audio input of the transceiver (Clover uses AFSK, not FSK). Receive audio is supplied to the controller from the external speaker jack or other source.

    • An IBM-PC computer or compatible. The computer must be at least a 286-level machine.

    • A P-38 Clover controller board or a DXP38 HF modem (see Chapter 2). All Clover products are available exclusively from HAL Communications (see the Resource Guide for details). The HAL P-38 Clover controller is installed *inside* the computer using any available expansion slot. The board uses a dual-microprocessor design and digital signal processing to achieve signal modulation and demodulation. The DXP38 HF modem, on the other hand, is an external device.

Clover-II Specifications for the HAL P-38 and DXP38 Processors

Format: 4-tone emission
Bandwidth: 500 Hz @ −50 dB
ARQ Mode Modulation: BPSM, QPSM, and 8PSM
FEC Mode Modulation: 2DPSM, BPSM, QPSM, and 8PSM
Tones: 2250 Hz Center
Symbol Rate: 31.25 per second
Error Correction Coding: Reed-Solomon code: 60, 75 or 90% code rate
ARQ Protocol: 2-level, multiblock auto adaptive modulation
Adaptive Mode Control: Measure S/N & Phase on all data, set TX Mode
ARQ Thru-put: 8 to 35 bytes/second
FEC Thru-put: 4 to 25 bytes/second
FSK RTTY Codes: Baudot and ASCII
ASCII Codes: 45, 50, 57, 75 baud
RTTY Tones: 1275/1445 or 2125/2295 Hz
AMTOR Code: CCIR-476 & CCIR-625
AMTOR Modes: ARQ or FEC
AMTOR Tones: 1275/1445 or 2125/2295 Hz
AMTOR Rate: 100 baud
PACTOR Modes: Auto-ARQ & FEC
PACTOR Rates: 100/200 baud

• HAL PC-Clover software. This is supplied by HAL Communications and is included with every P-38 and DXP38. It is *not* a terminal program. The PC-Clover software is the instruction set of the controller itself! It's loaded into the controller's memory each time you decide to operate. This approach makes it easy to update the controller in the future. You simply buy a new diskette or download the software. There are also third-party programs that will function with either product.

The original Clover controllers cost approximately $1,000. In the late '90s HAL introduced the P-38 with a selling price of less than $400. Then, in early 1999, the external DXP38 appeared, also selling at under $400. Both products are multimode processors. That is, they offer RTTY, AMTOR, PACTOR (HAL refers to it as "P-Mode") and Clover.

CLOVER ON THE AIR

Most Clover-equipped stations are dedicated to relaying high-volume message traffic, often functioning as HF/Internet e-mail gateways (see Chapter 9). This makes casual Clover operating the exception rather than the rule, although you will find occasional keyboard-to-keyboard "live" chats.

Clover signals are relatively easy to recognize. The data bursts vary in length. Some are short, while others can last several seconds. The signals make a staccato *brrrrr* sound rather than the chirping rhythms of PACTOR and G-TOR.

Clover Conversations

When it comes to on-the-air operating, Clover is different from any of the modes we've discussed so far. To call CQ, for example, you switch to the **MODE** menu, highlight **CQ** and press **ENTER**. The Clover controller sends a CW identification followed by a raucous stream of data. Unlike other digital modes, you do not see "CQ CQ CQ" flowing across your screen. In fact, you see nothing at all.

The controller sends CQ in the form of data signals that appear as CQ "flags" to other Clover stations. When another Clover operator tunes in your signal, all he sees is a statement on his screen announcing that you are calling CQ. At that point he can ignore you or press a single key to establish a Clover connection.

Once the conversation has started, you need to let the other station know when you've completed a statement. Remember that there are no *over* commands. For example:

QST Reviews the HAL Communications DXP38

(from April 2000 *QST*)

The HAL DXP38 does not offer as many modes as other MCPs, but it does extremely well with what it *does* offer: PACTOR, RTTY, ASCII, AMTOR and CLOVER II. The DXP38 manual is the best of the group. It is well written, well organized and concise. The writing style is conversational with a slight touch of humor. You hardly need the manual to install the DXP38. Unlike the other MCPs in this group, the DXP38 uses RCA phono jacks on the rear panel, which makes cabling a breeze. (If you've ever soldered several wires onto a DIN plug, you know what we mean.)

Once you're up and running, you have your choice of FSK for RTTY, AMTOR and PACTOR, or AFSK for the entire set. Clover II must be sent using AFSK, so we opted for AFSK for all of our testing. A trim pot to adjust the transmit audio level is accessible from the rear panel. HAL Communications has a long history in HF digital communication. Its RTTY terminal units are still considered among the best in the world. It's no surprise, then, that the RTTY performance of the DXP38 was the best in the group. In weak-signal conditions and brutal contest environments, the DXP38 consistently copied RTTY when the other MCPs displayed mostly gibberish.

The hardware tuning indicator—an LED emulation of the traditional "crossed bananas" oscilloscope display—was a joy to use. Even weak, interference-laden signals could be tuned quickly and accurately. The DXP38's PACTOR performance (HAL refers to this as *P-Mode*) was also outstanding. We were able to establish and hold PACTOR links under marginal conditions. Only the SCS PTC-IIe could top the DXP38 in this category.

We ran into difficulty testing the DXP38 on AMTOR, but that had nothing to do with the device itself. AMTOR signals are as rare as proverbial hen's teeth these days and the only way we could conduct AMTOR tests with this or any of the other MCPs was to arrange skeds. Despite the hassles, the DXP38 seemed to acquit itself very well in this mode. And then there is Clover II. This complicated 4-tone mode is the chief competitor to SCS's

PACTOR II. Both modes have been doing battle for dominance in the commercial and amateur markets for years. Of course, both companies insist that their mode is superior and can present evidence to prove their cases. In our brief, informal tests... it was a draw. We did not notice substantial performance differences between PACTOR II and Clover II. Under identical conditions both modes appeared to transfer our sample files in roughly the same amounts of time.

As with PACTOR II, the overall performance of Clover II was remarkable, maintaining links and transferring data in conditions under which we could just barely hear the other stations. Clover requires high transceiver stability ($\pm$ 5 Hz drift per hour, maximum) and slow, careful tuning for optimal results. The DXP38's tuning indicator was helpful, but it is only updated every 2 seconds in Clover II, so you have to tune very carefully. Once you find a Clover signal (it sounds like an extended *brrrrrrr*) and tune it in, the best thing to do is leave your radio alone. Even a slight VFO tweak is sufficient to break the link. HAL Communications was the only company to provide both *DOS* and *Windows* software with their product. We used the *Windows* software and it worked extremely well. (Some preferred the *Windows* tuning indicator for Clover II rather than the hardware display.) Unlike the other burst modes, Clover II is bi-directional, which means that it is sending and receiving information at regular intervals without waiting for an "over" command from the operator. This made live QSOs a bit tricky because the one operator could begin commenting on something you said before you were even finished saying it! Despite the excellent performance of Clover II, hams have not embraced this mode in large numbers. The fact that the Winlink 2000 e-mail network is almost exclusively based on PACTOR will not help this situation in the immediate future. Clover BBSs and live QSOs are not as scarce as G-TOR or AMTOR contacts, but they are not plentiful, either.

**Hello! My name is Steve
and I live in Wallingford, Connecticut. I am new to
Clover. What do you think of it? >>>**

Without >>>, BTU, K or a similar symbol at the end of my statement, the other operator might inadvertently jump in after the end of the first line. This can be very confusing for everyone!

While you're watching the conversation, it's easy to get distracted by the receive/transmit status table in the upper right corner of the screen (if you're using the HAL software). The table displays the modulation format in use at the moment, the signal-to-noise ratio, tuning error, phase dispersion, error-correction capacity and transmitter output power (as a percentage of full output). The table is split into horizontal rows labeled "MY" and "HIS." Not only do you see your own parameters changing, you see the changes taking place *at the other station*! (Clover accomplishes this feat by periodically swapping station data.) Who is enjoying the best receive conditions? Which station is doing the greatest amount of error correcting at the moment? Just look at the table!

CLOVER BANDWIDTH

As we've already discussed, space is a premium in the HF digital subbands. That's why it's important for any HF digital mode to be as *narrow* as possible.

With a 2-kHz bandwidth, an HF packet signal is nearly as wide as a voice transmission. You can't squeeze too many packet signals onto the band before serious interference begins (just listen to the HF packet activity on 20 meters!).

As remarkable as it may seem, Clover manages to conduct extremely efficient communications while using only 500 Hz of spectrum! *Four* Clover signals could fit in the same amount of spectrum required for a one HF packet signal.

This narrow bandwidth is yet another prominent feature of Clover. Of course, a narrow-bandwidth signal requires more careful tuning. You can also understand why your transceiver must be very stable to operate Clover. With a 500-Hz wide signal, all it takes is a little bit of drift and you're way out of the ballpark!

Chapter SEVEN

G-TOR

Like Clover and PACTOR II, G-TOR is a *proprietary* HF digital mode. In this case the Kantronics Corporation holds the rights to G-TOR and the mode is available *only* in Kantronics multimode communications processors, primarily the KAMPlus and KAM98 (see Chapter 2).

WHO PUT THE "G" IN G-TOR?

G-TOR is an acronym for *Golay*-coded *Teleprinting Over Radio*. Golay coding is the error-correction system created by M. J. E. Golay and used by the *Voyager* spacecraft. Sending billions of bytes of data across the Solar System required a scheme to ensure that the information could be recovered despite errors caused by interference, noise, and so on. If Golay coding could meet that challenge, the engineers at Kantronics wondered if it could be applied to digital communication on the HF bands as well.

To create G-TOR, Kantronics combined the Golay coding system with full-frame data interleaving, on-demand Huffman compression, run-length encoding, a variable data rate capability (100 to 300 bit/s) and 16-bit CRC error detection. G-TOR system timing is liberal enough to permit long-distance communication.

The G-TOR waveform consists of two phase-continuous tones (BFSK) spaced 200 Hz apart (Mark = 1600 Hz, Space = 1800 Hz). However, the system can still operate at the familiar 170-Hz shift (Mark = 2125 Hz, Space = 2295 Hz). The optimum spacing for 300-bit/s G-TOR transmissions would normally be 300 Hz. In the interest of keeping the bandwidth as close to 500 Hz as practical, some small amount of performance is traded off to save bandwidth.

One of the primary causes of reduced throughput on synchronous ARQ signals such as those used by PACTOR is errors in the acknowledgment signal (ACK). To reduce unnecessary retransmissions due to faulty ACKs, G-TOR uses *fuzzy* ACKs. This system allows receiving stations to tolerate a small number of errors in an ACK signal, rather than ignoring it completely and automatically resending the data.

With these innovations, G-TOR performs quite well despite noise, fading and interference. It is capable of passing information at two to three times the rate of PACTOR. Under some conditions it can even rival the performance of Clover and PACTOR II.

G-TOR IN ACTION

G-TOR is surprisingly easy to operate. If you're familiar with PACTOR operating, G-TOR is essentially the same. When you're in the G-TOR mode, you use FEC to call CQ. This allows your transmission to be copied by as many stations as possible. If someone wants to talk to you, they respond in G-TOR using your full call sign.

When you call CQ, it sometimes helps to let stations know that you're fishing for G-TOR contacts. Here's an example of a typical G-TOR CQ:

CQ CQ CQ CQ CQ—G-TOR
CQ CQ CQ CQ CQ—G-TOR
CQ CQ CQ CQ CQ—G-TOR
DE WB8IMY WB8IMY WB8IMY
Standing by for G-TOR calls. K K

Once the G-TOR link is established, the station that called you is the ISS, or *information sending station*. It's up to him to speak first. While he is transmitting, you're the *information receiving station*, or IRS. You flip these IRS/ISS roles back and forth by sending an "over" code in a manner similar to PACTOR. And like PACTOR, G-TOR can transfer ASCII or binary information.

If you copy someone calling CQ G-TOR, you simply switch to the G-TOR standby mode. Establishing a connection is as easy as entering the word G-TOR followed by the call sign of the station you wish to contact. For example:

GTOR WB8IMY

G-TOR Frame Structure and ARQ Cycle

G-TOR operates as a synchronous ARQ mode. Regardless of transmission rate, the cycle duration is always 2.4 seconds, data frames are 1.92 seconds long, and the acknowledgments take .16 seconds. At 300 baud, each data frame contains 69 bytes of data, one control byte, and a two byte CRC. At 200 baud the frame contains 45 data bytes, and at 100 baud 21 data bytes.

Synchronization is established during the linking phase when the calling station (master) sends a G-TOR frame containing TO and FROM call signs. The information receiving station (IRS), while in standby, synchronizes to the frame by searching for its call sign. Once in step, it acknowledges such to the master and sends <link established> to its terminal. Transmission of data may then begin. Sufficient time is left between the end of the data frame and the start of the acknowledgment for propagation between stations over an HF path. A change in information flow direction (changeover) is accomplished by extending the acknowledgment bytes into a changeover frame. Once acknowledged by the other station, changeover is complete. Link quality, dictated by the number of consecutive good or bad frames received, determines link speed. The effective performance of stations, while communicating over adverse HF channels, relies on the combined use of forward error correction, interleaving, and redundancy.

Prior to transmission, 300-baud frames are divided into 48 12-bit words and matched with 48 error correction words of 12 bits each. The entire 72 byte data frame is then interleaved bit by bit, resulting in 12 bins of 48 bits, and transmitted. Upon reception by the IRS, the reverse process is carried out. The frame is synchronized, de-interleaved, decoded, and checked for proper CRC. If the frame is found to be in error, the IRS will request that the matching parity frame be sent. Upon receipt, the parity frame is used in combination with the data frame in an attempt to recover the original data bits. If unsuccessful, the ARQ cycle begins again. The dispersement of noise-burst errors via interleaving and the power of the Golay code to correct 3 bits in every 24 usually results in the recovery of error-free frames.

Golay Error-Correction Coding And Interleaving

G-TOR uses extended Golay coding which is capable of correcting three or fewer errors in a received 24-bit code word. The Golay code used in G-TOR is a half-rate code, so that the encoder generates one error-correction bit (a parity bit) for every data transmitted. Interleaving is also used to correct burst errors which often occur from lightning, other noise, or interference. Interleaving is the last operation performed on the frame before transmission and de-interleaving is the first operation performed upon reception. Interleaving rearranges the bits in the frame so that long error bursts can be randomized when the de-interleaving is performed. When operating at 300 symbols/sec, the interleaver reads 12-bit words into registers by columns and reads 48-bit words out of the registers by rows. The de-interleaver performs the inverse, reading the received data bits into registers by rows and extracting the original data sequence by reading the columns. A long burst of errors, for example 12-bits in duration, will be distributed into 48 separate 12-bit words before the error correction process is applied. This effectively nullifies the errors. Both data frames and parity frames are completely interleaved. In addition, by using the invertibility characteristic of Golay code words, data frames are always alternated with data frames coded in Golay parity bits. In this way, G-TOR can maintain full speed (when band conditions are good)—rather than fall to rate-1/2—receiving parity bits that can be used as data or as parity.

You can monitor the G-TOR data rate by glancing at the front panel of the KAM. When the **STA** LED is off, the rate is 100 bit/s. A flashing **STA** indicates 200 bit/s and a steady **STA** means that you're perking along at 300 bit/s. The data rates change automatically based on the quality of the link. Links always begin at 100 bit/s. If the number of correctly received frames exceeds a preset value, the receiving station will request a speed increase to 200 or 300 bit/s. If the link deteriorates, the data rate will automatically ratchet downward. You can also set the data rate manually.

The G-TOR mode supports the KAM mailbox. If you set the **ARQBBS** command *ON*, any connecting station will receive the mailbox prompt. You'll also find some BBSs that accept G-TOR connects.

G-TOR'S FUTURE?

Despite G-TOR's outstanding performance, and the fact that it is available in almost every Kantronics KAM-series controller (thousands have been sold worldwide), the mode has yet to be embraced by large numbers of HF digital operators. The reasons for this relative lack of interest are complex.

G-TOR debuted in 1994, the year the Internet began to emerge as a powerful medium for global digital communication. Hams suddenly were not as keen on exploring new high-performance HF digital modes. Those who established HF gateways for Internet e-mail had more experience and confidence with Clover and PACTOR, so they tended to build their stations around those modes. Others who simply wanted to chat, contest and chase DX decided that RTTY (and later, PSK31) was adequate for the purpose. Despite its positive technical attributes, G-TOR suddenly found itself in the unfortunate position of being a mode without an application.

As this book was being written, G-TOR's high-performance competitors, Clover and PACTOR II, were appearing in more affordable devices. In fact, the cost of a Clover or PACTOR II-equipped multimode controller is only about $50 to $100 more than a Kantronics KAMPlus or KAM98. This puts G-TOR in head-to-head competition with Clover and PACTOR II. The market will ultimately decide its success or failure.

Chapter EIGHT

Hellschreiber

Are the Hellschreiber modes really digital? Or, are they a hybrid of the analog and digital worlds?

Some argue that the Hellschreiber modes are more closely related to facsimile since they display text on your computer screen in the form of images (not unlike the product of a fax machine). On the other hand, the elements of the Hellschreiber "image text" are transmitting using a strictly defined digital format rather than the various analog signals of true HF fax or SSTV.

The Hellschreiber concept itself is quite old, developed in the 1920s by Rudolf Hell. Hellschreiber was, in fact, the first successful direct printing text transmission system. The German Army used Hellschreiber for field communications in World War II and the mode was in use for commercial land-line service until about 1980.

As personal computers became ubiquitous tools in ham shacks throughout the world, interest in Hellschreiber as an HF mode increased. By the end of the 20th century amateurs had developed several sophisticated pieces of Hellschreiber software and had also expanded and improved the Hellschreiber system itself.

Like RTTY and PSK31, the Hellschreiber modes are intended for live conversations. Most of the activity is found on 20 meters, typically between 14.061 and 14.065 MHz.

FELD-HELL

Feld-Hell is the most popular Hellschreiber mode among HF digital

Murray, ZL1BPU, working a Japanese station on Feld-Hell over a 10,000 km path.

experimenters. It has its roots in the original Hellschreiber format, adapted slightly for ham use.

Each character of a Feld-Hell transmission is communicated as a series of dots, with the result looking a bit like the output from a dot-matrix printer. A key-down state is used to indicate the black area of text and the key up state is used to indicate blank or white spaces. One hundred and fifty characters are transmitted every minute. Each character takes 400 ms to complete. Because there are 49 pixels per character, each pixel is 8.163 ms long.

Feld-Hell characters can be sent by using a simple CW transmitter, but most operators prefer to achieve the same effect by feeding 900 or 980-Hz tones an SSB transmitter. Whichever method is used, the timing requirements are precise. Fortunately, Rudolf Hell developed a simple technique of printing the text *twice* to deal with the effects of phase shifts and small timing errors. (The text is not transmitted twice, however.) In this sense you could say that Feld-Hell is a *quasi-synchronous* mode. The font was designed so that the top and bottom of each line of text could be matched, if necessary, to create readable words, no matter what phase relationship existed between transmitting and receiving equipment.

Equipping a Feld-Hell Station

If you already own an SSB transceiver, you can try Feld-Hell by building a Hamcomm modem such as the one described in Chapter 2. You can find Feld-Hell software that will work with this simple modem, or you can bypass the external modem approach completely by using soundcard-based software. See the Resources section of this book.

Stability is important to prevent display distortion, but that is the only strict requirement of a Feld-Hell transceiver. Feld-Hell is not a burst mode like PACTOR, Clover or G-TOR, so you don't need to be concerned about transmit/receive switching speeds. Almost any SSB transceiver can be used for Feld-Hell work.

EA2BAJ running 16-tone concurrent MT-Hell.

MT-HELL

MT-Hell or *Multi-Tone* Hell transmissions are similar in concept to Feld-Hell, but rather than using on/off keying to represent the black and white pixels that make up the text, MT-Hell uses *frequency variations.* As with Feld-Hell, all you need is a reasonably stable SSB transceiver to operate MT-Hell. The modulation and demodulation can be performed with a soundcard, Hamcomm-type interface or a Motorola DSP development kit. See the Resources section of this book.

Shifting audio tones are used to designate the black/white pixel elements, and each row of pixels is sent using a different frequency. The DSP software at the receiving end only detects the *presence* of the MT-Hell signaling frequencies—it doesn't care about their shapes or amplitudes. Since noise and interference are amplitude-modulated in nature, the software can effectively "ignore" the garbage. The result is excellent performance under marginal conditions.

The data rate of MT-Hell is not fixed, because there is no attempt to achieve any sort of synchronization. The only timing constraints are those that ensure that the transmitted characters are displayed in the proper order and with minimal distortion. Using slow transmission rates and DSP detection software, the weak-signal performance can be improved even further.

There are at least four variations of MT-Hell in use today:

C/MT-Hell, or *Concurrent MT-Hell,* uses many tones (seven or

A test transmission on 80 meters between ZL1AN and ZL1BPU using 7-tone sequential MT-Hell.

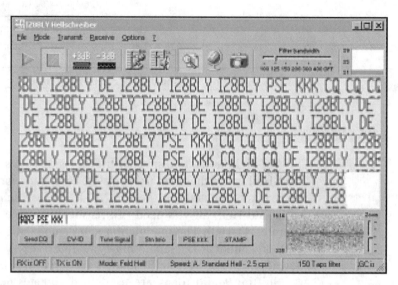

Windows software developed by IZ8BLY allows you to transmit and receive Feld-Hell, several MT-Hell modes and Morse using your computer soundcard.

more), usually transmitted at the same time. It is the most popular of the MT-Hell modes.

S/MT-Hell or *Sequential MT-Hell* was invented in 1998 by ZL1BPU. It uses only a few tones, typically five or seven, but never more than one at a time.

FSK-Hell is a two-tone system commonly using 980 Hz for black and 1225 Hz for white with a 245-Hz shift. You'll find it included in *Windows*-based software developed by IZ8BLY. G3PPT's *FELDNEW8* also offers FSK-Hell. Both programs operate at 122.5 baud.

Duplo-Hell is a relatively new variant invented by IZ8BLY. The font and format are identical to Feld-Hell, except that two columns are transmitted at the same time, using two on-off keyed tones. The tones and shift are the same as FSK-Hell.

Chapter NINE

MFSK

(This chapter contains excerpts from "MFSK for the New Millennium" by Murray Greenman, ZL1BPU, which appeared in the January 2001 issue of QST.*)*

MFSK is really a type of super-RTTY. The MFSK technique was developed during the heyday of teleprinter HF communications as a way to combat multipath propagation problems, providing reliable point-to-point communications with relatively simple equipment. Piccolo, for example, was used on diplomatic links between England and Singapore, and typically provided good copy for an hour after RTTY links had faded out. The technology was then electromechanical, but several very important principles were recognized at the time, and maximum advantage taken of them:

• The performance (reduced error rate) improved as the number of tones used increased.

• The performance was best when the least number of symbols is used to represent each transmitted text element.

•With a special integrating detector, tones as closely spaced as the baud rate could be uniquely detected without cross-talk.

Piccolo and Coquelet both used two symbols per text character—compare this with 7.5 for RTTY and anywhere from three to twelve for PSK31. MFSK16 uses only one symbol per signaling element! With MFSK modes the baud rate (rate at which symbols are transmitted) is rather lower than the text rate. This is because each symbol carries more information in its frequency properties than RTTY or PSK. While this is confusing, this technique has the advantage that the longer symbols are easier to detect in noise, have a lower bandwidth, and are much less

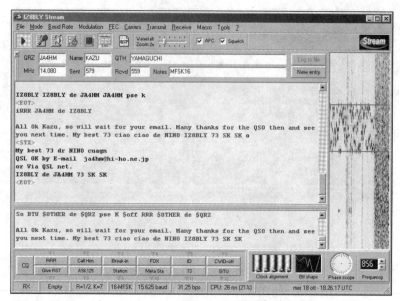

The MFSK *Stream* software for *Windows* developed by Nino Porcino,
IZ8BLY.

affected by multipath timing errors.

Piccolo originally used as many as 32 tones, but the most common form used six. Coquelet generally used 12 tones. MFSK has been recently tested with as many as 64 tones, although the released version, MFSK16, uses 16 tones, and the weak signal variant MFSK8, uses 32.

The integrating detector used in Piccolo was a milestone in FSK detection techniques in its day. Without going into great detail, narrow active filters, with very high gain, were used to detect each tone. By carefully choosing the baud rate and tone channel spacing, and resetting the filters at the start of each symbol period, it was possible to reliably detect very weak tones without crosstalk to adjacent channels, and in fact the response of the adjacent channels produced a null at the point of sampling. This not only assisted with noise rejection, but prevented energy resulting from ionospheric effects on one tone from appearing in the next channel.

THE NEW APPROACH

In searching for a better way to hold reliable long path QSOs, Murray Greenman, ZL1BPU, looked at what made copy difficult with existing

An MFSK16 spectrogram. (The horizontal lines are 300 Hz apart.)

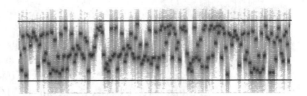

modes, and what could be done about it. It was obvious that Phase Shift Keying, unless relatively high speed, was not practical. The incidental phase errors introduced by an unstable ionosphere (particularly in polar regions) typically exceed the phase modulation of the signal. Frequency Shift Keying and On-Off Keying also perform poorly, but principally because the arrival time of signals vary, often 5-10 ms, depending on the path, and perhaps by as much as 30 ms between long and short paths. This is longer than a 22 ms RTTY symbol, and multipath reception is the reason why so many RTTY signals, even of good signal strength, will not print reliably. So, in casting around for some better method, the old MFSK techniques were revisited. In association with this research, modern PC and soundcard DSP techniques were reviewed. Putting these together, it was found that all the necessary building blocks were available to not only replicate, but enhance, the old MFSK modes, using nothing more than a PC with a soundcard.

Murray decided to kick things off by sending a specification for the new mode to many DSP, coding and software experts, and a remarkable collection of ideas and offers of assistance resulted. Nino Porcino, IZ8BLY, of Hellschreiber and MT63 fame, quickly turned the specification into reality. The result has been tested thoroughly under both real and simulated conditions. The very first QSO using this new mode was over 11,000 miles, long path on 17 meters, with 100% copy using 25 watts and dipole antennas, so the specification can't have been far wrong!

The first QSO was between IZ8BLY and ZL1BPU on June 18, 2000, and since then these two have been in communication using this mode almost every day. Most days they work 20 meters or 17 meters, using as little as 5 W.

Like IZ8BLY Hellschreiber and the well known PSK31 programs, all you need to run this software is a Pentium class PC with a soundcard, and a couple of simple cables. The first software for MFSK16 is called *Stream*, and is available for downloading on the Web. See the Resources section of this book.

THE SIGNAL

What does this new mode consist of? Well, there are 16 tones, sent one at a time at 15.625 baud, and they are spaced only 15.625 Hz apart. Each tone represents four binary bits of data. The transmission is 316 Hz wide, and has a CCIR specification of 316HF1B. It is exactly like RTTY, but with 16 closely spaced tones instead of two wider spaced tones. With a bandwidth of 316 Hz, the signal easily fits through a narrow CW filter.

The tones are continuous phase keyed, which eliminates keying noise, and the phase information can be used to determine tuning and symbol phase.

Unlike Piccolo or even PSK31, no special arrangements are made to transmit symbol timing, since this can be recovered from the inherent properties of the signal. A most important factor is that like RTTY, the signal is constant amplitude—it does not require a linear transmitter to maintain a clean signal. Driving the transmitter too hard on MFSK16 will not make the signal any wider.

To ensure that text is received with an absolute minimum of errors, the new mode utilizes an excellent Forward Error Correction (FEC) technique, using Viterbi decoder routines by Phil Karn, KA9Q, and a clever self-synchronizing interleaver developed for MFSK by IZ8BLY. The typing rate, even with FEC, is over 40 WPM. This speed is achieved by efficient coding techniques, including a varicode similar to PSK31, providing an extended ASCII character set.

Finally, the receiver detector uses a synchronous Fast Fourier Transform (FFT) routine, a DSP technique which exactly models the original Piccolo integrating detector. The FFT is also able to provide phase information, AFC and a "waterfall" tuning display. The filter provides 4 Hz wide channels, and is easily able to separate the 16 closely spaced tones.

The signal has an amusing musical sound, is quite narrow, clean to tune across, and not unpleasant to listen to. The sound is certainly better and the bandwidth narrower than some modes on HF these days!

FIRST IMPRESSIONS

Downloading the *Stream* software and installing it is very simple. Fortunately the help information is also available as a separate download, so you can read that before you install.

Stream offers a generous collection of tools along the top, separate transmit and receive windows, a good collection of definable "macro" buttons, and an excellent "waterfall" tuning display. Along the bottom is a list of settings and parameters, plus the date and time. There is also a drop-down log window, for automatic logging and insertion of QSO information, and a very useful "QSP" window for relaying incoming text.

Nino's software actually includes three modes! The default mode is MFSK16 (16 tone 16 baud MFSK with FEC), and there is also slower but more sensitive MFSK8 (32 tones 8 baud with FEC). Both modes have the same bandwidth, just over 300 Hz, but sound quite different. The other new mode is one of Nino's own, 63PSKF, which is a 63 baud PSK mode, like

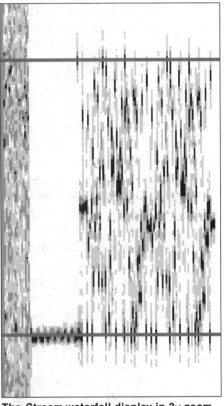

The *Stream* waterfall display in 3× zoom mode.

PSK31 but faster, and with full-time FEC. PSK63F is about 100 Hz bandwidth. The MFSK and PSK modes are complementary, as Nino's new mode is great for short path DX and local QSOs. You'll have no trouble telling them apart, and no trouble telling Nino's PSK63F from PSK31, because it is twice as wide. As a standard of comparison, the software includes PSK31 as well!

The software is very simple to use—start typing and it transmits, and press F12 to end the transmission. The trouble comes in tuning in the MFSK signals. It takes some skill and a certain patience learning to tune in MFSK, but the results are worth the effort.

Because the tones are closely spaced and the filters very narrow, you must have a very stable transceiver, and you must use the tuning provided with the software, not the transceiver tuning, and certainly not the RIT! The software allows you to tune up and down in 1 Hz steps, or click on the waterfall for exact tuning. The waterfall has a zoom function, and zoom ×3 is best.

The AFC is good, but you need to be within about 5 Hz of the right place to start with. The AFC works on the idle tone, which appears often enough for the AFC to catch it. You can see this happen as the Phase Scope comes alive. You can also manually tune by clicking on the waterfall display in just the right spot, or use the Up/Down Frequency buttons to tweak the tuning.

Tuning is done using the waterfall display. Under the lower horizontal line (red on the screen) you'll see a broad band towards the left. This is the idle carrier, the lowest of the 16 tones. This carrier is transmitted briefly at the start of each over, and returns at the end, or whenever the operator stops to think. All you need to do is center the red line on this carrier, and the AFC will keep it there. During the over, you'll see little black vertical stripes all over the waterfall, with gray "side-lobes" above and below. These are the transmitted symbols, and once again, you can adjust the software tuning so the red line centers the lowest of these symbols. Unfortunately while this is easy when the signal is already tuned, finding the correct spot on a weak signal during an over is not so simple and takes a little practice.

Once you've found the right spot, almost perfect text will start to appear on the screen, although it is delayed by some 3-4 seconds as the data trickles through the error correction system, and appears one or two words at a time.

The mode is a delight to use once you learn to tune in. The typing speed is fast, and while changeover from transmit to receive is not as fast as RTTY or Hellschreiber, it is quite good enough for ragchewing and nets.

PERFORMANCE

For QSOs using short path, say distances to 8,000 miles on 20 meters not on polar routes, MFSK16 works fine, but you may find PSK31 easier to use. If you are interested in QRP, there is not much to choose between the two. On long path, over the poles, and in really difficult conditions where QRM or instability is the major problem, MFSK16 is out on its own. It keeps going, giving almost perfect copy, without slowing down, when the signal is barely audible, has bad fades, has noticeable Doppler, multipath and even QRM. High power isn't necessary.

MFSK16 is also probably the best mode yet for digital work on the lower bands. If you are into traffic handling, or sending bulletins on 80 or 40 meters, give this mode a try. It just does not give up! 80 meters is especially prone to multipath, as RTTY and Feld-Hell users know. MFSK16 will work thousands of miles on 80 meters at night, running one watt, and with 90% perfect copy! Lightning is largely ignored.

Although not noticeably better on the low bands, MFSK8 is great to have when the band starts to die. It is definitely more sensitive than MFSK16, and although tuning is very tight, and typing speed down to 25 WPM, it will allow you to complete that QSO with almost perfect copy.

The other mode offered in the *Stream* package, PSK63F, is quite the reverse. Although not very good on long path, it is sensitive, almost as good as PSK31, and is fast (40 WPM). It uses FEC to give 100% error-free copy most of the time. It is also very easy to tune, since it is wider than PSK31 and includes excellent AFC. It is also much less affected by Doppler and drift problems. PSK63F makes a good short-haul DX mode, and would be good on VHF.

MFSK Technical Specifications

By Murray Greenman, ZL1BPU

A. Objectives
A.1. Use
An easy to use Amateur chat mode for real-time QSOs, nets and broadcast of bulletins, but not intended for contesting or BBS use. A half-duplex non-ARQ forward error corrected mode.

A.2. Slick Operation
The default mode offers comfortable typing speed and quick manual half-duplex end of over turnaround, for reasonably slick operation. Turnaround time will not exceed five seconds using the default mode.

Comment: Slick operation requires both low latency (the time for data to trickle through the system), and low turnaround time (the time to end an over plus the time to start a transmission and have synchronism acquired at the other end). Obviously the typing speed is dependent on mode and FEC settings, and the turnaround time will scale with baud rate changes and increase with addition of FEC.

Very slick operation (as required for RTTY and CW contests) is not considered to be a goal of this mode. Latency is not an issue at all for beacon and bulletin broadcast use.

A.3. DX Performance
Very good performance for long-haul DX, such as long path 20 meters. The software will be robust in fading conditions, and in moderate QRM such as is prevalent in Europe. The software will be robust in lightning and man-made burst noise such as is common on 80 meters.

A.4. Ease of Use
Operator ease of installation and use is considered to be very important. No special equipment will be required to set up a station to operate this mode.

A.5. LF and Super-DX
Paragraph deleted in this release.

B. Description of Modes
B.1. Choice of Modes
The choice of useful modes offered by the software is limited. Other modes offered will be sufficiently different in nature and application to be obvious to the non-technical user (for example PSK31 is a suitable alternative).

The default mode will be 16-FSK, 16 tone 15.625 baud with FEC ON, Interleaver ON. The FEC will be R=1/2 K=7, using the NASA algorithms. (CCIR definition 316HF1B)

B.2. Mode Names
The default mode will be known as **MFSK16**.

B.3. Mode Selection

The user software will automatically select number of tones, baud rate, FEC and interleaver regimes from a single menu selection, invisible to the user. The user will not be able to change any of these parameters, nor will they be shown what the settings are. See paragraph B.1.

Different modes with clearly different performance will only be provided from a limited list which descriptively names the modes.

C. Transmission Technique

C.1. MFSK Technique

The transmission will be based on 16-FSK (sequential single tone FSK), with continuous phase (CPSK) tones. There will be no delay between tones, and no shaping of the tones.

C.2. Performance Options

Paragraph deleted in this release.

C.3. Default Mode

Paragraph deleted in this release.

C.4. Transmission Bandwidth

Transmission bandwidth will be less than the **number of tones** × **tone spacing** × **2** at the −30 dB point relative to a single carrier when properly adjusted. The transmitter does not need to be linear.

Comment: This is outside the second sidelobe, so should be easily achievable. The CCIR requirement is for less than 0.005% of the total power to be outside the *necessary bandwidth*, which is 316 Hz for the default mode 16FSK16. See FCC Part 47, paragraph 2.202.

C.5. Symbol Rate and Tone Spacing

The system will use a tone spacing numerically equal to the baud rate. Each symbol will consist of a single square keyed pulse with the same start and finish phase as all others, and concatenated with others. No single or isolated pulses, or gaps between pulses will be emitted.

Comment: If *tone spacing = baud rate,* i.e. *spacing = 1/T*, orthogonal reception with non-coherent demodulation is assured.

C.6. Symbol Rates Offered

Paragraph deleted in this release. Only one symbol rate (15.625 baud) is offered.

C.7. Range of Tones

Transmission tones (and receiver tuning) should be adjustable to suit different transceiver IF filters. Low (~1kHz) and high (~2 kHz) alternatives to be accommodated without change of actual tone spacing.

C.8. Bit Stream Orientation

At the lowest level, i.e. the single symbol, the system will be a bitstream oriented transmission, allowing convolutional code FEC, Varicode and binary transfer options to be used at a higher level when specified.

Comment: This approach allows maximum flexibility; for example, it would also allow data block transmission over sequential tones, such as is used in Piccolo, with two sequential tones per character.

C.9. FEC Coding

Full-time FEC coding with interleaver will be used with this mode. FEC will be sequential R=1/2, K=7, using NASA algorithms.

The interleaver will be self-synchronizing, based on 10 concatenated 4×4 bit IZ8BLY diagonal interleavers.

C.10. Alphabet Coding

The default alphabet coding will be the IZ8BLY Varicode, using an extended ASCII character set and super-ASCII control codes.

C.11. Idle Limitation

To avoid extended periods of single tone (for example, when the keyboard buffer is empty), a non-printing character will be stuffed periodically into the transmit system as though it came from the keyboard. A non-printing character will be sent whenever the transmitter is idle for 20 symbol periods. Stuffing will not occur when the buffer is not empty.

Idle will be achieved by sending ASCII NULL or other non-printing character, followed by an extended zero bit stream, e.g. "0000000000000000." The latter will be rejected by the receiver as an invalid character.

Comment: It is intended in future releases of this specification to describe a "secondary data channel." Special characters for secondary channel data, which will be outside the extended ASCII character set defined in paragraph C.10, will allow transmission of low priority data during idle, exactly like the IZ8BLY MT63 "secondary channel." The data rate would be very low, but would permit automatic ID. These super-ASCII characters would be used in place of the above mentioned NULL character.

The purpose of the "diddle" is to allow the receiver symbol clock to remain in lock. The diddle must not be continuous, as the idle period is used for signal tuning purposes.

C.12. Beginning and End of Transmission

At the start of transmission an idle carrier representing the lowest tone will be transmitted for 8 symbol periods. At end of transmission all pending transmit characters will be sent, and the transmitter will be flushed with zeros, allowing an idle period of at least 4 symbol periods to be transmitted.

Comment: The purpose of the idle carrier is to allow accurate manual tuning.

C.13. Tone Bit Weighting

Tone weighting will be such that the lowest audio tone represents zeros in all bits. The weighting will increase in gray-code as the tone frequency is increased. This technique provides the least Hamming distance between adjacent tones:

Tone	Weight	Tone	Weight
0 (lowest)	0000	8	1100
1	0001	9	1101
2	0011	10	1111
3	0010	11	1110
4	0110	12	1010
5	0111	13	1011
6	0101	14	1001
7	0100	15 (highest)	1000

D. The Receiver

D.1. Demodulator Technique

The receiver will use non-coherent demodulation, using an FFT filter and demodulator technique, integrating the signal over the symbol tone period by sampling the period synchronously with the transmitted symbol. A recovered symbol clock will be used to accomplish this purpose.

Comment: Reduction of multi-path reception effects can be achieved by windowing the symbol sample period to exclude, for example, the first and last 5 ms worth of samples.

D.2. Decision Decoding

The symbol decision decoder which follows the FFT will preferably use bit-wise soft decisions. The symbol decision decoder may offer soft decisions to the FEC decoder.

D.3. Tuning Indicators and AFC

The symbol decision decoder will provide tuning indication, as well as signal performance metering (S/N meter). Automatic Frequency Control may be offered.

Comment: This mode will be very sensitive, and very narrow, so will require very accurate tuning. Due consideration needs to be given to accurately tuning almost inaudible signals, through use of a good expanded waterfall display. A S/N meter in arbitrary units based on data from the symbol decision decoder will be useful. AGC is not required with an FFT approach.

D.4. FEC Decoding

The FEC decoder will use soft decisions, and may use soft data from the symbol decoder.

Comment: The FEC regime can identify the FEC dibits unambiguously from the fixed weight bits in the received symbols.

D.5. Confidence Meter

The FEC decoder may provide information for a "signal confidence" meter, which reflects the current reliability of FEC decoding from the decoder metrics.

D.6. Appropriate Settings

Paragraph deleted in this release.

D.7. Default Text Mode

The default receiver text mode will use a specially adapted varicode, as defined in section C.10., translating into ASCII for display. Characters outside the defined extended ASCII character set will not be displayed, i.e. will not be sent to the screen.

Comment: This use of non-printing special characters permits these characters to provide symbol sync or carry a secondary channel text thread where supported by the software, without requiring support to be part of this specification.

E. Other Requirements

E.1. Symbol Sync

Receiver symbol synchronization will be affected by using transmitted carrier phase or data transition information, or by multiple FEC decoder voting, rather than recovering AM modulation, since the transmitter is hard keyed.

Comment: The original Piccolo used AM modulation of tones to transmit sync. MFSK16 does not, but transmits constant-phase carriers. CPFSK transmissions make these techniques possible.

E.2. Symbol Clock

The received symbol phase will remain substantially correct (within 90 degrees) for at least 50 symbol clock periods while receiving an idle condition, and will retain phase with one non-printing character every 20 symbols.

E.3. Tuning Display and Tuning

The user interface will provide a waterfall tuning display, preferably with point and click tuning, but with at least easy tuning in steps of 1/4 of the tone spacing or better.

E.4. User Interface Windows

The user interface will provide separate TX and RX windows, a type-ahead buffer and simple macros.

E.5. Mode Menu

See paragraph B3.

E.6. Identification and Tuning Signal

Morse ID and a transmitter tuning signal will be provided.

Suggestion: OOK of the lowest tone, or FSK between lowest and highest tones would be suitable for the ID, as it would assist receiver tuning. Transmitter idle will suffice for the tuning signal.

E.7. Transmitter Control

Transmit control will be via VOX or serial port, using the WD5GNR standard.

E.8. Receiver Squelch

Receiver squelch may be offered as an option. It may be based on Symbol Decoder metrics.

E.9. USB/LSB Reverse Switch

A reverse switch may be provided to allow copy of wrong-sideband signals. If provided, the transmitter will reverse in concert with the receiver.

F. Documentation

F.1. Software Copyright

User software (released binaries) of software offering this mode will be copyright of the author, and all rights will be retained by the various authors. All source code copyright to be retained by the author. All source code must be submitted to the MFSK development team if the software claims to be written to this specification. The development team "committee" will review all source and executables before approval is given. See section F.3.

Submitted source code will not be published without the authors' permission, but if the software is approved will be retained and offered to interested developers on request.

Permission to use the MFSK16 logo will be provided for incorporation into approved software.

F.2. Non-Amateur Use

Commercial or military use of the released binaries will be permitted only as agreed by the contributing authors and owner of this specification. No responsibility for damages or lack of performance is assumed by the owner, the authors, or any of the development team.

Comment: The purposes of this paragraph are to (a) discourage commercial exploitation of the product or concept without approval and involvement of the developers; and (b) to protect the developers from unreasonable demands.

F.3. Publication of Specifications

Algorithms and code tables sufficient to define the mode will be published as electronic documents, are referred to and referenced in this document. The documents will be retained on the MFSK Web site at **www.qsl.net/zl1bpu/MFSK**.

A reference version of the source code will be made available to interested developers. These released documents, along with this specification, will constitute the complete description of the mode.

F.4. User Documentation

Software developers are to make their own arrangements for documentation of their software. Installation and operating information will accompany each release.

F.5. Release of Non-compliant Software

No user software will be released which purports to support this specification but which offers for public use MFSK operation not covered by this specification. This particularly applies to test versions and beta versions. Test features in these releases must be protected from public access.

F.6. Freeware Availability

All user software (released binaries) will be made available free of charge, other than media costs, perhaps via an end-user license. The software will be fully operational and fully documented.

Chapter TEN

WinLink 2000: The HF E-Mail Connection

The Internet has become the e-mail medium of choice for most hams. But there is a sizeable group of amateurs who often travel beyond the reach of the Internet. This group includes hams at sea, travelers in recreational vehicles (RVs), missionaries, scientists and explorers. No doubt the day will come when wireless, affordable Internet e-mail access will be available from any point on the globe. Until that day arrives, however, the Amateur Radio HF digital network is a very capable substitute!

THE EVOLUTION OF WINLINK 2000

More than 30 HF digital stations worldwide have formed a remarkably efficient e-mail network. Running *WinLink* software and using PACTOR or PACTOR II, these facilities transfer e-mail between HF stations and the Internet. They also share messages between themselves using Internet forwarding.

The network evolved in the 1990s from the original AMTOR-based *APLink* system. APLink was a network of stations that relayed messages to and from the VHF packet network. As PCs became more powerful, and as PACTOR and Clover superseded AMTOR, a new software system was needed. That need brought about the debut of *WinLink*, originally authored by Victor D. Poor, W5SSM, with additions from Peter Schultz, TY1PS. *WinLink* itself evolved with substantial enhancements courtesy of Hans Kessler, N8PGR. To bring the Internet into the picture *WinLink* stations needed an e-mail "agent" to interface with cyberspace. To meet that requirement Jim Jennings, W5EUT, added *NetLink*. Early in 2000 the system took the next evolutionary leap, becoming a full-featured Internet-to-HF gateway system known as WinLink 2000.

Thanks to these advancements, an HF digital operator at sea, for example, can now connect to a WinLink 2000 station and exchange Internet e-mail with nonham friends and family. He can also exchange messages with other amateurs by using the WinLink 2000 station as a traditional "mailbox" operation.

WinLink 2000 stations scan a variety of HF digital frequencies on a regular basis, listening on each frequency for about two seconds. By scanning through frequencies on several bands, the WinLink 2000 stations can be accessed on whichever band is available to you at the time.

WinLink 2000 features include…

• Text-based e-mail with binary attachments such as DOC, RTF, XLS, JPG, TIF, GIF, BMP, etc.

• Position reporting and inquiry accessible from both the Internet via APRS, e-mail or radio to track the mobile user.

• Graphic and text-based weather downloads from an extensive list of weather products.

• Pickup and delivery of e-mail regardless of the participating station accessed.

• End-user control of which services and file sizes are transmitted from the participating stations.

• The ability for each user to re-direct incoming e-mail messages to an alternate e-mail address.

• The ability to use any Web browser to pick up or deliver mail over WinLink 2000.

HARDWARE REQUIREMENTS

In addition to your HF SSB transceiver, you'll need a multimode processor that is capable of communicating in binary mode with either PACTOR I or PACTOR II. At the time this book was published, only five processors met this requirement:

• Kantronics KAM+ or KAM98
• SCS PTC-II or PTC-IIe
• Timewave/AEA PK-232

ACCESSING A WINLINK 2000 STATION

I strongly recommend, as a new user, that you use *AirMail* software to streamline the message exchange process. *AirMail* is free for downloading from several sites on the Internet. See the Resources section of this book.

If you're already set up to operate PACTOR or PACTOR II, you can connect to a WinLink 2000 station right away. Just choose a station, dial in the frequency and transmit the connect request, along with the proper call sign. A list of WinLink 2000 stations is shown in **Table 10-1**. A continually updated list is maintained on the Web at **winlink.org/k4cjx/**.

Remember that WinLink 2000 stations usually scan through several frequencies. If you can't seem to connect, the WinLink 2000 station may already be busy with another user, or propagation conditions may not be favorable on the frequency you've chosen. Either try again later, or try to connect on another band.

Sending E-mail To and From the Internet

From the Internet side of WinLink 2000, friends and family can send e-mail to you just as they would send e-mail to anyone else on the Net. In fact, the idea of WinLink 2000 is to make HF e-mail exchanges look essentially

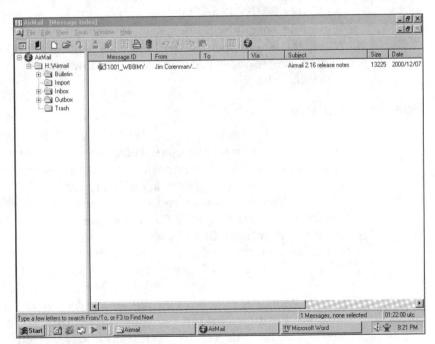

AirMail software for *Windows* makes it easy to send and receive e-mail through the WinLink 2000 network.

the same as regular Internet e-mail from the user's point of view.

Internet users simply address their messages to **<your call sign>@winlink.org**. For example, a message addressed to **wb8imy@winlink.org** will be available to me when I check into *any* WinLink 2000 station.

Through the *AirMail* software I can address messages to non-hams, or to other hams, for that matter, by using a similar format. Here are some examples:

SP BillGates@microsoft.com (This command will send an Internet e-mail to my friend Bill Gates at Microsoft.)

SP N1RL@winlink.org (This will send a message to N1RL, who is also able to access the WinLink 2000 network from HF.)

SP W1NRG@K1RO.CT.NA (I can also send messages to hams who can access VHF/UHF packet. Note that I needed to include the complete routing in the destination address.)

Although WinLink 2000 supports file attachments, remember that the radio link is *very* slow (especially compared to the Internet). Sending an attachment of more than 25,000 bytes is usually not a good idea. DOC, XLS, JPG, BMP, GIF and TIF files are permitted as long as they are small enough to comply with the particular user-set limit. Where possible, WinLink 2000 compresses files. For example, a 400,000 byte BMP file may be compressed to under 15,000 bytes. However, this is an exception rather than the rule.

Position Reports

Another fascinating feature of WinLink 2000 is the ability to provide position reports for mobile users. There are two ways in which a family member or friend may query position reports and determine where you are (or at least where you *were* at the last report).

The first method of obtaining a position report from a mobile user is to request it from the Automatic Position Reporting System, APRS. Simply input the URL **winlink.findu.com/(call)** into your Web Browser where (call) is the actual amateur call sign of the specific station you are looking for. Successive map views will appear identifying the location of the station.

The second method is to send an e-mail message addressed to **qth@winlink.org** requesting the report. The subject line of the message must be "**Position Request**" (without the parentheses). The body of the message contains the call letters of interest (one call letter per line).

For example:

To: **qth@winlink.org**
Subject: **Position Request**

W4NWD <== Message Body
KG8OP 3
KF4TVU 999
REPLY: SMTP:jblow@yahoo.com

In the above example, for the entries in the position reports table for W4NWD, only the *most recent* report will be sent back. No number after the call letters is the default. For KG8OP, the *last 3* most recent reports will be returned, and for KF4TVU, 999 will return *all* the position reports for KF4TVU. In addition to the originator of the position request message, a reply will also be sent to e-mail address *jblow@yahoo.com*, and W5CTX, a radio user. Again, in all cases a reply will be sent to the originating e-mail address. Be patient when you send requests; it can take about 15 minutes to receive a reply.

Here is an example of the requested position report:

Automated Reply Message from Winlink 2000 Position Report Processor
Processed: 12/17/1999 19:02:55

w4nwd Sat 11 Dec 1999 10:30 UTC 10-40.87N 061-38.03W

Powerboats on hard, Trinidad, W.I , 1Coat Antifoul on !
Received by Winlink: Sat 30 Dec 1899 00:00 UTC

kg8op Tue 14 Dec 1999 14:48 UTC 25-20.20N 077-49.17W 5.8 KTS 133 CRS

Underway to Nassau from Chub Cay, Berry Islands, Bahamas
Received by Winlink: Tue 14 Dec 1999 15:01 UTC

kg8op Mon 13 Dec 1999 13:44 UTC 25-34.95N 078-31.02W

Chub Cay, Berry Islands, Bahamas
Received by Winlink: Sun 12 Dec 1999 12:58 UTC

kg8op Sun 12 Dec 1999 13:04 UTC 25-34.95N 078-31.02W

Chub Cay, Berry Islands, Bahamas
Received by Winlink: Sun 12 Dec 1999 12:58 UTC

kf4tvu Wed 15 Dec 1999 10:44 UTC 26-46.99N 080-00.63W 5.7 KTS 030 CRS

Underway to St Lucie inlet; local time = 05:44 am
Received by Winlink: Wed 15 Dec 1999 12:12 UTC

kf4tvu Wed 15 Dec 1999 03:40 UTC 26-45.85N 080-02.61W 0.0 KTS 226 CRS

anchored for night in Lake Worth
Received by Winlink: Wed 15 Dec 1999 03:49 UTC

Table 10-1

WinLink 2000 Stations (as of January 2001)

All stations shown are auto-replying in either Pactor-1 or Pactor-2 and, unless noted, are available 24 hours a day. All frequencies shown in kilohertz.

Explanation of codes:
Antennas
B = Beam
D = Dipole
V = Vertical
Internet details
E = E-mail address sysop
H = Homepage URL
I = Internet Telnet TCP/IP ID

Region 1—Africa and Europe

F6CDD.#31.FMLR.FRA.EU, Rene, Toulouse
D 3578.9 3580.9 3583.9 3585.9 7036.9 7037.9 7039.9
B 14069.9 14073.9 21072.9 28101.9
E f6cdd@onetelnet.fr
I 44.151.31.25
OH2NPE.FIN.EU, Teo, Espoo
B 14073.9
E teo.jasterberg@pp.inet.fi
SM6USU.SWE.EU, Svante, Glose
D day: 3589.9
B night: 14075.9
E jabina@swipnet.se
ZS5S.ZAF.AF, Joost, near Durban
B 7037.0 7041.0 14064.0 14069.0 14073.0
21064.0 21069.0 21073.0 28069.0 28073.0
AF/EU/NA: 0800-1400 (28/21) 1500-0700 (28/21/14/7)
AS: 0700-0800 (28/21) 1400-1500 (21/14)
E zs5s@zs5s.com
H zs5s.com or users.iafrica.com/z/zs/zs5s

Region 3—Asia and Oceania

HS0AC.#PWT.BKK.THA.AS, Rudolf/DL1ZAV,
Bangkok
14069.4 14072.4
0800-1600z: 21072.4 28169.4
E kruggelr@siemens.co.th
VK7PU.#BUR.TAS.AUS.OC, Phil, Burnie
D 2000-2300: 7032.9 7036.9 7039.9 7040.9 7042.9 7044.9
B 2300-1000: NE - 1000-2000: EU (SP)
14067.9 14070.9 14074.9 14076.9 14078.9 14079.9
E philharb@southcom.com.au
ZL2UT.NZL.OC, Basil, Gisborne
B East-2400-0800: 14074.9, 1800-2400: 28097.9
E b.davoren@clear.net.nz

Region 2—Americas

K4CJX.#MIDTN.TN.USA.NA, Steve, Nashville
D 3618.9 3621.9 7072.4 7073.9 7076.9
10121.9 10122.9 10123.9 10125.9
14064.4 14065.9 14073.9 14076.9 18103.9
E k4cjx@home.com
H winlink.org/k4cjx
K6IXA.#CCA.CA.USA.NA, Grady, Atwater
B 120°: 14064.9
E gradyw@elite.net
K7AAE.WA.USA.NA, Ron, near Seattle
3619.9 7070.9 10126.9 14074.9 21071.9
E k7aae@yahoo.com
KA6IQA.#SAN.CA.USA.NA, Tom, near San Diego
3619.9 7069.9 10125.9 14063.9 18102.9 21069.9
E lafleur@ucsd.edu
KB6YNO.ME.USA.NA, Eric, South Portland
V 0100-1130: 7071.9, 1130-0100: 14067.9
E kb6yno@maine.rr.com
H www.qsl.net/kb6yno
KF6NPC.#SOCA.CA.USA.NA, Mike/N6KZB, Riverside
D 1912.9 3615.9 3623.9 7067.9 7076.9

10125.9 10137.9
 14066.9 14077.9 18105.9 21069.9 24925.9
28076.9 28176.9
E n6kzb@arrl.net
KN6KB.#MLBFL.FL.USA.NA, Rick, nr
Melbourne/FL
V 3616.9 3617.9 7068.9 7104.4 10122.9 10125.9
 14066.9 14067.9 21077.9 28072.9
E rmuething@cfl.rr.com
H www.dwatt.com/users/kn6kb.html
N0ZO.#LAKE.FL.USA.NA, Pat, Lady Lake
V 7065.9 7070.4 7071.9 7076.9 7101.4
 10121.9 10123.9 10125.9 10127.9 10140.4
10142.9
 14066.9 14068.9 14112.4 14117.9
 1300-2300: 18104.9 21072.9 28072.9
E n0zo@lcia.com
H members.lcia.com/n0zo
N8PGR.#NEOH.OH.USA.NA, Hans, near
Cleveland
D H24: 7065.9 7067.9 7071.9 7101.4 7104.4
 10127.4 10137.9 10140.4 14074.9 14117.9
18101.4
 2200-1400: 3620.9 3621.9 3622.9 3629.9
 1200-0300: 14067.9 14072.9 14075.9 14077.9
 1200-2000: 21072.9 21074.9 21079.9 28074.9"
28079.9
E n8pgr@custsys.com
VE1YZ.#HFX.NS.CAN.NA, Neil, near Halifax NS
D 3619.9 3621.9 7071.9 7074.9 7076.9
V 10123.9 10125.9 10127.9
V 14068.9 14072.9 14112.4 14117.9
B South: 21074.9 28076.9
E ve1yz@fox.nstn.ca
W6IM.USA.NA, Rod, San Diego Yacht Club
B 135° - 7073.9, 10136.9, 14071.9
E mclennan@home.com
W7BO.#SWWA.WA.USA.NA, John, near
Portland
D 0400-1500: 3620.9 3621.9 7069.9 7071.9
B 14.065.9 14073.9 14075.9 21072.9 28076.9
 1600-1900 NA/CA, 1800-2330 EU, 2330-1600
OC
E jburke@worldaccessnet.com
H www.qsl.net/w7bo

W9GSS.IL.USA.NA, Chuck, Peoria
36239.9 7072.9 10125.9 14073.9
E w9gss@home.com
W9MR.#SEIL.IL.USA.NA, Ken, Keensburg
V: 3620.9 3621.9 7063.9 7065.9 7067.9
 10125.9 10140.4 10142.4
 1300-0500: 14065.9 14107.9 14113.9
 1400-0400: 18101.9 21061.9 24922.9 28074.9
E w9mr@midwest.net
WA6OYC.NC.CA.USA.NA, Oakland Yacht Club
- Alameda
 V 3626.9 3627.9 7068.9 7069.9 10129.9
10131.9 14072.9 14075.9
18102.9 18105.9 21065.9 21067.9 24925.9
24927.9
28074.9 28076.9 28176.9
 E petelismer@aol.com
WB0TAX.#SHRV.LA.USA.NA, Deni, Shreve-
port
D 3619.9 3623.9 7067.9 7070.4 7101.4
 10125.9 10127.9 10129.9 18104.9
B 14066.9 14068.9 21072.9 21074.9 28072.9
E deni@dwatt.com
H www.dwatt.com
WB5KSD.#ETX.TX.USA.NA, Jon, nr Dallas,
(00JUL/K4CJX)
D 3620.9 3622.9 3624.9 7072.9 7074.9 7076.9
7101.4
 10127.9 10132.9 10133.9 10141.4 14068.9
14072.9 14074.9 14113.9
E jonbrown@hyperusa.com
WG3G.TYCMM.TTO.CAR.SA, Bernie, Trinidad
 1230-1030: 3617.9 7036.9 7100.9 10125.9
10142.9
 14062.9 14066.9 14109.9 18102.9 21079.9
24920.9 28070.9
E wg3g@wow.net
ZF1GC.#GC.CYM.CAR.NA, Frank, Grand
Cayman Is
D 7071.9 7072.9 10123.9 10125.9 10140.9
B 14063.4 14065.9 14073.9 14075.9 14076.9
18100.9
 21063.9 21065.9 21066.9 28073.9
E zf1gc@candw.ky

Table 10-2
Common WinLink 2000 Mailbox Commands

The first column lists the primary command (short form). You may optionally spell out commands as listed in the second column (alias).

Cmd	Alias	Function
A	ABORT	Sending station shall Stop Sending
B	BYE	Log off from WinLink
CANCEL #		Delete my message number # (1)
D #	DELETE	Delete my message number# (1)
E	EXPERT	Expert Mode (ON/OFF); Show current mode w/o parameters
H	HELP	Read the Main Help file
K #	KILL	Delete my message number # (1)
I	INFO	Read the Information File
L	LIST	List all my PRIVATE and NTS messages
LM	LIST	List all my PRIVATE and NTS messages
LB		List new Bulletins since your last inquiry
LB #		List Bulletins in the system, starting with Bul #
LLB # (loc)		List last # Bulletins filed, beginning at "loc" (2)
LH		List all HELP messages/files
LL		List new Messages since your last inquiry (2)
LL # (loc)		List last # messages filed, beginning at "loc" (2)
LN		List all my unread messages
LOGOUT		Log out from WinLink
LR		List users in last 24 hours
LT		List all NTS messages in the system
LY		List this station's usage statistics
L> CALL LTO		List all messages addressed to "CALL" 2 (1)
L< CALL LFM		List all messages from "CALL" (2)
L@ CALL		List all messages addressed "@ CALL" (2)

Cmd	Alias	Function
NTS		List all "pending" NTS messages
PAGE		Page Mode (ON/OFF).
		Show current mode w/o parameter
R #	READ #	Read a specific message number WITHOUT headers
RC		Read all my pending messages without confirmation
RH #		Read a specific message number WITH routing headers
RM	RN	Read all my unread messages WITHOUT headers
RMM # # ...		Read multiple messages
RMMC # # ...		Read multiple messages without confirmation
SP CALL	S	Send a PRIVATE message to CALL. (3)
SP CALL@BBSCALL		Send a PRIVATE message to CALL at BBSCALL (4)
SB TOPIC		Send a BULLETIN to TOPIC
SB TOPIC@DESTN		Send a Bulletin to TOPIC at DESTINATION
SFM		Send a file of contiguous messages
ST		Send a NTS message
TIME		Show daily BBS usage limit and time left
T	TALK	Alert the Sysop
V	VERSION	Ask for the current WinLink software version

Messages are terminated with the /EX character sequence starting at the beginning of a new line and followed by an <ENTER>.

Notes:

(1) You may still re-read a Deleted message until the next MBO message purge cycle, but you must remember its message number.

(2) MBO users can see any message in the system while keyboarders can only see messages addressed TO or FROM themselves. (loc) allows you to start the list with that message number, the oldest message (O) or the newest message (N).

(3) Sending a message without routing information may stay at that MBO until the recipient connects to that same station to read it.

Chapter ELEVEN

HF Digital Contesting

Nothing tests your equipment and operating skills like a contest. If you have shortcomings in your antenna system, you're going to discover them very quickly. If your transceiver can't seem to handle the crunch of dozens of signals in proximity, it will become painfully apparent within minutes. If you need better logging or contest software, the first hour of a contest will provide a powerful motivation to upgrade!

Some hams shun contesting because they assume they don't have the time or hardware necessary to win—and they are probably right. But winning is *not* the objective for most contesters. You enter a contest to do the best you can, to push yourself and your station to whatever limits you wish. The satisfaction at the end of a contest comes from the knowledge that you were part of the glorious frenzy, and that you gave it your best shot!

Contesting also has a practical benefit. If you're an award chaser, you can work many desirable stations during an active contest. During the ARRL RTTY Roundup, for example, some hams have worked enough international stations to earn a RTTY DXCC. Jump into the North American RTTY QSO Party and you stand a good chance of earning your Worked All States in a single weekend.

Always remember that contesting is ultimately about *enjoyment*. The thrill of the contest chase gets your heart pumping. The little triumphs, like working that distant station even though he was deep in the noise, will bring smiles to your face.

You'll make a number of friends along the way, too. Why do you think they call it the "brotherhood of contesting"?

DIGITAL CONTESTING TIPS FOR THE "LITTLE PISTOL"

If you are like most amateurs, your station falls into the Little Pistol category. Mine does, too. The Little Pistol station usually includes a low- to medium-priced 100-W transceiver feeding an omnidirectional antenna such as a dipole, vertical, end-fed wire and so on. Even if you are blessed with a beam antenna on a tower or rooftop, in all likelihood you are still a Little Pistol.

At the opposite end of the spectrum are the Big Gun stations. They

Digital Contest Calendar

See *QST* or the *National Contest Journal* for complete rules.

Month	Date	Contest
January	1st	SARTG New Years RTTY Contest
January	First full weekend	ARRL RTTY Roundup
January	Last full weekend	BARTG RTTY Sprint
February	First full weekend	NW QRP Club Digital Contest
February	First full weekend	FMRE RTTY Contest
February	Second weekend	WorldWide RTTY WPX Contest
March	First full weekend	Ukraine RTTY Championship
March	Second weekend	North American RTTY Sprint
March	Third weekend	BARTG WW RTTY Contest
April	First full weekend	EA WW RTTY Contest
April	Third weekend	TARA PSK31 Rumble
April	Fourth full weekend	SP DX RTTY Contest
May	First full weekend	ARI International DX Contest
May	Second weekend	VOLTA WW RTTY Contest

have multiple beams on tall towers, phased array antennas on the lower bands, incredibly long Beverage receiving antennas and more. They own high-end (read: expensive) transceivers that boast razor-sharp selectivity and they usually couple them to 1.5 kW linear amplifiers.

So how does a Little Pistol go about beating a Big Gun in a digital contest? With rare exceptions, he can't. Banging heads with a Big Gun station is usually an exercise in futility. He is going to hear more stations, and more stations are going to hear him. Focus, instead, on using your wits and equipment to get the best score possible. You probably won't beat a Big Gun, but you could make him sweat!

Month	Date	Contest
June	Second weekend	ANARTS WW RTTY Contest
July	Third weekend	North American RTTY QSO Party
July	Fourth full weekend	Russian WW RTTY Contest
August	Third weekend	SARTG WW RTTY Contest
August	Last full weekend	SCC RTTY Championship
September	First full weekend	CCCC PSK31 Contest
September	Last full weekend	CQ/RJ WW RTTY Contest
October	First full weekend	TARA PSK31 Rumble
October	First full weekend	DARC International Hellschreiber Contest
October	Second Thursday	Internet RTTY Sprints
October	Third weekend	JARTS WW RTTY Contest
November	Second weekend	WAE RTTY Contest
December	First full weekend	TARA RTTY Sprints
December	Second weekend	OK RTTY DX Contest

Check Your Antenna

How long has it been since you've inspected your antenna? Is everything clean and tight? And have you considered trying a different antenna? Contests are terrific testing environments for antenna designs. For example, if you've been using a dipole for a while, why not try a loop? Just set up the antenna temporarily for the duration of the contest and see what happens.

Check Your Equipment

If you are having problems with RF getting into your processor or soundcard, fix these glitches *before* you find yourself in a contest! Also, consider spending $100 or so to install a 500-Hz IF filter in your radio if it doesn't have one already. Outboard audio DSP filters are helpful to cut down the signal clutter, but a good IF filter will make a world of difference in a contest environment. Finally, check your computer and software. Make sure you understand the program completely. Set up any "canned" messages/responses and test them before the contest begins.

Understand Propagation

If you are participating in a DX contest, is it a good idea to prowl the 40-meter band in the middle of the day? No, it isn't. Forty meters is only good for DX contacts after sundown. If you are hunting Europeans or Africans on 10 or 15 meters, when is the best time to look for them? The answer is late morning to early afternoon.

In other words, let propagation conditions guide your contest strategy. Your goal is to squeeze the most out of every band at the right time of day. For example, I might start the ARRL RTTY Roundup on 10 meters and then bounce between 10, 15 and 20 meters until sunset. After sundown, I'll concentrate on 20 and 40 meters. In the late night and the wee hours of the morning, I'll probably limit myself to 40 and 80 meters.

In some contests you can enter as a single-band station. Which band should you choose? That depends on the contest and your equipment. If you have a 15-meter beam and you are entering a DX RTTY contest, it makes sense to concentrate on 15 meters, even though you'll probably find yourself with nothing to do in the late evening after the band shuts down. Twenty meters makes sense as the ultimate "around the clock" band, but it is often very crowded during contests, and populated with Big Gun stations. In some instances you may find it more "profitable" to concentrate your efforts on another band without so many signals.

Hunting and Pouncing vs. "Running"

The common sense rule of thumb is that a Little Pistol station should only hunt and pounce. That means that you patrol the bands, watching for "CQ CONTEST" on your monitor, and pouncing on any signals you find. The Big Guns, on the other hand, often set up shop on clear frequencies and start blasting CQs. If conditions are favorable, they'll be hauling in contacts like a commercial fishing trawler! This is known as *running*.

In many cases Little Pistols are probably wasting valuable time by attempting to run. There are situations, however, where running *does* make sense. If you've pounced on every signal you can find on a particular band, try sending a number of CQs yourself to catch some of the other pouncers. If you send five CQs in a row and no one responds, don't bother to continue. Move to another band and resume pouncing.

Seek Ye the Multipliers

Every contest has multipliers. These are US states, DXCC countries, ARRL sections, grid squares and so on, depending on the rules of the contest. A multiplier is valuable because it multiplies your total score.

Let's say that DXCC countries are multipliers for our hypothetical contest. You've amassed a total of 200 points so far, and in doing so you made contacts with 50 different DXCC countries.

$200 \times 50 = 10,000$ points

Those 50 multipliers made a huge difference in your score! Imagine what the score would have been if you had only worked 10 multipliers?

If given a choice between chasing a station that won't give me a new multiplier and pursuing one that *will*, I'll spend much more time trying to bag the new multiplier.

Choose the Right Mode

Many contests allow the use of all HF digital modes, but RTTY is by far the most popular. Yes, you can probably make some contacts during the ARRL RTTY Roundup using PACTOR, for example, but you'll miss RTTY contact points while doing so. As the saying goes, "When in Rome do as the Romans do."

There are some contests that are devoted to one mode in particular. With the increasing popularity of PSK31, for example, PSK31-*only* contests are springing up. Watch for the TARA PSK31 Rumbles in April and October, and the CCCC PSK31 contest in September. In October you'll find the DARC Hellschreiber contest (see the sidebar).

The DARC International Hellschreiber Contest

The International Hellschreiber Contest is sponsored by the Deutscher Amateur Radio Club (DARC) to increase interest in the Hell modes.

- **Dates:** The first full weekend of October. Saturday: 80 meters, 1400-1600 UTC. Sunday: 40 meters, 0900-1100 UTC.
- **Entry categories:** Single operator, short wave listener (SWL)
- **Contacts:** The same station may be contacted once on each band for QSO and multiplier credit.
- **Exchange:** Signal report and QSO number starting with 001.
- **Points:** One point per completed QSO. Passing traffic (QTC) counts for additional points. QTCs can be reported back during a QSO with another station. A QSO may be reported only once and not back to the originating station. A QTC contains the time, call sign, and QSO number of the station being reported (1214 / DF5BX / 003). A maximum of five QTCs can be sent to or received by the same station, which can be worked several times to complete this quota.
- **Multipliers:** DXCC countries and all Japanese, US and Canadian call districts.
- **Logs:** Please send a separate log for each band. Logs must contain: Band, Date, Time UTC, Call sign, Message sent and received, Name, QTCs, Points and Multipliers.
- **Scoring:** Sum of the total points and QTCs × total multipliers. Log must be sent within four weeks after the end of the contest to: Werner Ludwig, DF5BX, Postfach 1270, D-49110 Georgsmarienhuette, Germany.

CONTEST SOFTWARE

No one says you have to use software to keep track of your contest contacts, but it certainly makes life easier! One of the fundamental elements of any contest program is the ability to check for duplicate contacts or *dupes*. Working the same station that you just worked an hour ago is not only embarrassing, it is a waste of time. The better contest

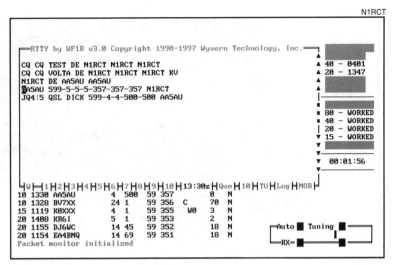

RTTY by Wyvern Technologies is a full-featured contest program that runs under DOS.

programs feature automatic dupe checking. When you enter the call sign in the logging window, the software instantly checks your log and warns you if the contact qualifies as a dupe. The more sophisticated programs "know" the rules of all the popular digital contests and they can quickly determine whether a contact is truly a dupe under the rules of the contest in question. Some contests, for example, allow you to work stations only *once*, regardless of the band. Other contests will allow you to work stations once per band.

A good software package will also help you track multipliers. It will display a list of multipliers you've worked, or show the ones you still need to find.

Any of the contest software packages advertised in *QST* and the *National Contest Journal* will function well. To minimize headaches, however, my suggestion is to stick with software written specifically for digital contesting. There are three popular digital contesting software packages on the market today:

• *RTTY* by WF1B is an excellent *DOS* package that is compatible with almost any multimode communication processor. *RTTY* is not a mere logging program; it is a terminal program that sends and receives data from a processor or modem. *RTTY* checks for dupes and "flags" them in

Ham System is a *DOS* digital contest package popular with European amateurs.

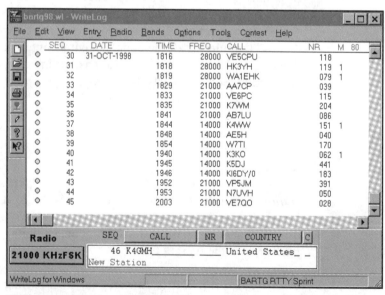

WriteLog is *Windows* software that not only logs, it includes the ability to send and receive RTTY and PSK31 using your computer soundcard.

color on your screen. It prepares your log for electronic submission. *RTTY* can also control your radio, changing bands and frequencies on the fly. *RTTY* will run on 486-level PCs and will also operate in a *DOS* window under *Windows 95* and *98*. You'll find complete information about *RTTY* on the Web at **www.wf1b.com/**.

• OH2GI *Ham System* is another *DOS* package similar to *RTTY* in many respects. *Ham System* is particularly popular among European contesters and is compatible with almost every known modem and processor. See the *Ham System* Web site at **www.kolumbus.fi/ jukka.kallio/**.

• *WriteLog* by Ron Stailey, K5DJ, is a full-featured *Windows*-based software package with an interesting twist. While *WriteLog* can be used with any processor or modem, it also has the *built-in* ability to send and receive RTTY and PSK31 using your soundcard. Like *RTTY*, *WriteLog* offers automatic dupe checking and flagging, multiplier displays, radio control, canned messages and more. See the *WriteLog* site on the Web at **www.writelog.com/**.

Chapter TWELVE

Resources

MULTIMODE COMMUNICATION PROCESSORS

Kantronics, 1202 East 23rd St, Lawrence, KS 66046; tel 785-842-7745; **www.kantronics.com**

MFJ Enterprises, PO Box 494, Mississippi State, MS 39762; tel 601-323-5869; **www.mfjenterprises.com/**

HAL Communications, 1201 W. Kenyon Rd, Urbana, IL 61801-0365; tel 217-367-7373; **www.halcomm.com/**

Timewave Technology Inc, 58 Plato Blvd E., St Paul, MN 55107; 651-222-4858; **www.timewave.com/**

SCS, Roentgenstr 36, D-63454 Hanau, Germany; **www. scs-ptc.com/**

HF DIGITAL MODEMS

Tigertronics, Box 5210, Grants Pass, OR 97527; tel 541-474-6700 **www.tigertronics.com**

MFJ Enterprises, PO Box 494, Mississippi State, MS 39762; tel 601-323-5869; **www.mfjenterprises.com/**

SOFTWARE

A few words about Web addresses: Every attempt was made to ensure that the following Web addresses were active at the time this edition was published. Web addresses change with astonishing frequency, however, so don't be suprised if a particular address is suddenly "broken." The person who manages the site may have removed it from the Web, or the site may have been moved to another address. When you encounter a faulty address, try using Web search engines such as AltaVista, Yahoo, Lycos and others to locate the site.

AirMail: A *Windows* program for sending and receiving messages via Amateur Radio PACTOR stations on the WinLink network. *AirMail* was specifically written for the SCS PTC-II and PTC-IIe PACTOR II controllers, but the latest versions support the Kantronics KAM-series controllers as well. **www.airmail2000.com**

BlasterTeletype: RTTY with your sound card. **www.geocities.com/SiliconValley/Heights/4477/**

DigiPan: PSK31 with your soundcard. **members.home.com/hteller/digipan/**

DSP-CW: CW and RTTY with your sound card. **www.zicom.se/dsp/index.html**

EasyTerm: A popular *Windows* terminal program for Timewave, HAL and Kantronics processors. **www.tiac.net/users/henley/eztpage.html**

HamComm: Allows you to communicate in RTTY, AMTOR and a few other digital modes when used with a simple modem (see Chapter 2). **www.pervisell.com/ham/hc1.htm**

Hellschreiber: *IZ8BLY Hellschreiber*, **iz8bly.sysonline.it**

Intercom: RTTY software for the Hamcomm modem (see Chapter 2). **ourworld.compuserve.com/homepages/pa3byz/rttymade.htm**

KA-Gold: High performance terminal software for Kantronics and TimeWave processors. **www.interflex.com/private/frame.htm**

Mix32W: Windows-based RTTY and PSK31 with your soundcard. **tav.kiev.ua/~nick/my_ham_soft.htm**

MMTTY: A sound-card-based RTTY system. **www.geocities.com/mmtty_rtty/**

Multimode: RTTY and PSK31 for Macintosh computers. PowerPCs recommended. **www.blackcatsystems.com/software/multimode.php3**

PacTerm: Terminal software for the KAM multimode processor. Creative services also supplies software for Timewave, MFJ, HAL and SCS products. **www.cssincorp.com**.

PSK31: For Linux, **aintel.bi.ehu.es/psk31.html**

Stream: MFSK for sound cards. **iz8bly.sysonline.it**

WinLink: A *Windows* program for MBOs which allows simultaneous operation on VHF packet and HF PACTOR and Clover (with suitable controllers), and provides store-and-forward message handling. **www.winlink.org/**

WinPSK: PSK31 for sound cards. **www.winpskse.com**

THE FCC OPENS THE DOOR IN 1995

Prior to 1995, the only legal HF digital modes were those that used standard ASCII codes. That severely limited the development of new modes that depended on more efficient coding. In 1995 the FCC declared that any new codes were legal as long as they were published in the public domain (and as long as the new modes that resulted did not otherwise violate Part 97 rules).

What follows is the actual FCC report. Although it discusses CLOVER, G-TOR and PACTOR specifically, this announcement opened the door to greater digital innovations.

FEDERAL COMMUNICATIONS COMMISSION
47 CFR Part 97
[DA 95-2106]
Use of CLOVER, G-TOR, and PacTOR Digital Codes
AGENCY: Federal Communications Commission.
ACTION: Final rule; interpretation.

SUMMARY: On October 2, 1995, the Chief, Wireless Telecommunications

Bureau adopted an Order that clarified that amateur stations may use any digital code that has its technical characteristics publicly documented. The amendments were necessary because some amateur operators have expressed concern about the propriety of using the CLOVER, G-TOR, and PacTOR codes on the High Frequency amateur service bands.

EFFECTIVE DATE: November 1, 1995.

FOR FURTHER INFORMATION CONTACT:

William T. Cross of the Wireless Telecommunications Bureau at (202) 418-0680.

SUPPLEMENTARY INFORMATION:

Order

Adopted: October 2, 1995

Released: October 11, 1995

By the Chief, Wireless Telecommunications Bureau:

1. This Order amends Section 97.309(a) of the Commission's Rules, 47 CFR 97.309(a), to clarify that amateur stations may use any digital code that has its technical characteristics publicly documented. This action was initiated by a letter from the American Radio Relay League, Inc. (ARRL).

2. The ARRL states that some amateur operators have expressed concern about the propriety of using the CLOVER, G-TOR, and PacTOR codes on the High Frequency (HF) amateur service bands. [CLOVER, G-TOR, and PacTOR are different techniques currently used by many amateur operators to increase the efficiency of digital communications transmitted on the HF portion of the radio spectrum.] This is due to the fact that Section 97.309(a) appears to authorize only the Baudot, AMTOR, and ASCII codes on the HF bands. [On the Very High Frequency and shorter wavelength bands, the rules authorize the use of any unspecified digital code provided the emission does not exceed a specified bandwidth. See Sections 97.307(f) (5)-(7) of the Commission's Rules, 47 CFR Secs. 97.307(f) (5)-(7).] The ARRL states that it has worked with the developers of CLOVER, G-TOR, and PacTOR to document the technical characteristics of these codes. It requests, therefore, that we amend Section 97.309(a) of the Commission's Rules to specifically authorize CLOVER, G-TOR, and PacTOR to remove any doubt about the permissibility of their use.

3. The primary purpose of CLOVER, G-TOR, and PacTOR is to facilitate communications using already-authorized digital codes, emission types, and frequency bands. The technical characteristics of CLOVER, G-TOR, and PacTOR have been documented publicly for use by amateur operators, and commercial products are readily available that facilitate the transmission and reception of communications incorporating these codes. [See Technical Descriptions CLOVER, G-TOR, PACTOR, published by the American Radio Relay League, Inc. (1995).] Including CLOVER, G-TOR, and PacTOR in the rules will not conflict with our objective of preventing the use of codes or ciphers intended to obscure the meaning of the communication. [The HF bands are widely used for international communications. Number 2732 Sec. 2.(1) of Article 32 Section I of the International Telecommunications Union Radio Regulations requires that transmissions between amateur stations of different countries by made in plain language. Section 97.113(a)(4) of the Commission's Rules, 47 CFR Sec. 97.113(a)(4), therefore, prohibits amateur stations from transmitting messages in codes or ciphers intended to obscure the

meaning thereof.] We agree, therefore, that it would be helpful to the amateur service community for the rules to specifically authorize amateur stations to transmit messages and data using these and similar digital codes. Accordingly, we are amending Section 97.309(a) to clarify the rules as requested by the ARRL.

4. Because the rule amendment adopted herein is interpretative in nature, and clarifies the existing amateur service rules, the notice and comment provisions of Section 553(b) of the Administrative Procedure Act, 5 U.S.C. Sec. 553(b), do not apply, and it is not subject to the publication or service requirements of Section 553(d) of the Administrative Procedure Act, 5 U.S.C. Sec. 553(d).

5. We certify that the Regulatory Flexibility Act of 1980 does not apply to the amended rule because there will not be any significant economic impact on a substantial number of small business entities, as defined by Section 601(3) of the Regulatory Flexibility Act. The amateur service may not be used to transmit communications for compensation, for the pecuniary benefit of the station control operator or the station control operator's employer, or for communications, on a regular basis, which could reasonably be furnished through other radio services. See 47 CFR Sec. 97.113. The Secretary shall send a copy of this Order, including the certification, to the Chief Counsel for Advocacy of the Small Business Administration in accordance with paragraph 605(b) of the Regulatory Flexibility Act, Pub. L. No. 96-354, 94 Stat. 1164, 5 U.S.C. Secs. 601-612 (1981).

6. Accordingly, IT IS ORDERED that effective upon publication in the Federal Register, Part 97 of the Commission's Rules, 47 CFR Part 97, IS AMENDED as set forth below. This action is taken under the authority delegated to the Chief, Wireless Telecommunications Bureau, in section 0.331(a)(1) of the Commission's Rules, 47 CFR Sec. 0.331(a)(1).

Federal Communications Commission

Regina M. Keeney,
Chief, Wireless Telecommunications Bureau.

Rule Changes

Part 97 of Chapter I of Title 47 of the Code of Federal Regulations is amended as follows:

PART 97—AMATEUR RADIO SERVICE

1. The authority citation for Part 97 continues to read as follows:

Authority: 48 Stat. 1066, 1082, as amended; 47 U.S.C. 154, 303. Interpret or apply 48 Stat. 1064-1068, 1081-1105, as amended; 47 U.S.C. 151-155, 301-609, unless otherwise noted.

2. Section 97.309 is amended by adding paragraph (a)(4) to read as follows:
Sec. 97.309 RTTY and data emission codes.

(a) * * *

(4) An amateur station transmitting a RTTY or data emission using a digital code specified in this paragraph may use any technique whose technical characteristics have been documented publicly, such as CLOVER, G-TOR, or PacTOR, for the purpose of facilitating communications.

Appendix

Technical Descriptions

In 1989, the League issued a challenge to amateurs to develop ways of improving the performance of packet radio in the high frequency bands. Instead of responding with improvements to AX.25 packet radio, designers came up with different strategies to provide for efficient HF data communication technologies, namely PACTOR, PACTOR II, CLOVER, G-TOR, PSK31 and, most recently, CLOVER-2000. These initiatives by individuals and design teams in the United States, Germany and the UK quickly gained acceptance. Products and software supporting these technologies are currently available and in use by radio amateurs worldwide. These systems have found applications by other radio services seeking improvements in HF communications efficiency.

The technical descriptions that follow authored by Steven Karty, N5SK, in collaboration with Dennis Bodson, W4PWF, are intended to document the manifest technical characteristics of these systems. They are not intended as complete system definitions.

These systems carry no specific endorsement by the League except to applaud the innovative work by their developers. We encourage further study and development of HF digital communications in the amateur services.

CLOVER

1. INTRODUCTION

CLOVER is a digital communications mode that conveys 8-bit digital data over narrow-band high-frequency radio. It can also transfer ASCII text and executable computer files without using the additional control characters required in other digital modes, which decrease throughput. It measures signal conditions, and automatically changes modulation format and data throughput to match current link quality. Reed-Solomon data encoding provides forward error correction (FEC) within each data block to repair many errors without the need for retransmission.

2. WAVEFORM

The CLOVER waveform consists of four tone pulses, each of which is 125 Hz wide, spaced at 125 Hz centers. The four tone pulses are sequential, with only one tone being present at any instant and each tone lasting 8 ms. Each frame consists of four tone pulses lasting a total of 32 ms, so the base modulation rate of a CLOVER signal is always 31.25 symbols per second. Data is conveyed by changing the phase and/or amplitude of successive pulses at the same frequency. These changes are made only at the instants midway between the peaks of two successive pulses when their amplitudes are zero. The measured CLOVER modulation spectra is tightly confined within a 500 Hz bandwidth, with outside edges suppressed 50 dB to prevent interference to adjacent frequencies. Unlike other modulation schemes, the CLOVER modulation spectra is the same for all modulation formats. Additional key parameters of CLOVER modulation include a symbol rate of 31.25 symbol/s (regardless of the type of modulation being used), 2:1 voltage (6 dB power) crest factor, and a ITU-R emission designator of 500H J2 DEN or 500H J2 BEN.

3. TRANSMISSION PROTOCOLS

CLOVER normally operates over half-duplex links and uses a Reed-Solornon algorithm to provide FEC. This FEC may be used alone in FEC mode or combined with an Automatic Repeat reQuest (ARQ) protocol in ARQ mode to acknowledge each individual data block. The ARQ mode provides an effective adaptive control system which constantly measures the signal-to-noise ratio, frequency offset, phase dispersion, and errors on each block of received data. CLOVER evaluates these measured parameters and selects the best modulation format for the respective propagation conditions. The receive modem sends commands to the transmitting modem indicating which modulation format should be used for the next transmission. This process ensures selection of an optimum modulation format, allowing CLOVER to operate even with multipath and other HF propagation impairments.

4. REED-SOLOMON ERROR DETECTION AND CORRECTION ALGORITHM

Reed-Solomon FEC is used in all CLOVER modes. This is a powerful byte and block oriented error-correction technique, not available in other common HF data modes, and it can allow the receiving station to correct errors without requiring a repeat transmission. Errors are detected on octets of data rather than on the individual bits themselves. This error correction technique is ideally suited for HF use in which errors due to fades or interferences are often "bursty" (short-lived) but cause total destruction of a number of sequential data bits. Error correction at the receiver is determined by "check" bytes which are inserted in each block by the transmitter. The receiver uses these check bytes to reconstruct data which has been damaged during transmission. The capacity of the error corrector to fix errors is limited and set by how many check bytes are sent per block.

Check bytes are also "overhead" on the signal and their addition effectively reduces the efficiency and therefore the "throughput rate" at which user data is passed between transmitter and receiver. Efficiencies of 60%, 75%, or 90% can be invoked by using successively lower levels of Reed-Solomon encoding for error correction, or 100% efficiency by bypassing this algorithm. Better propagation conditions do not require as much error correction, which means the amount of overhead decreases and the efficiency increases.

5. SIGNAL CHARACTERISTICS

CLOVER normally uses six different modulation formats which are automatically selected to best compensate for the propagation conditions. These formats are described in the following chart:

CLOVER MODULATION FORMATS

Modulation	Description	Data Rate (bps)
2DPSM	Dual-diversity Binary Phase Shift Modulation (PSM)	62.5
BPSM	Binary PSM	125
QPSM	Quadrinary PSM	250
8PSM	8-level PSM	375
SPSM/2ASM	8-level PSM, 2-level Amplitude Shift Modulation	500
16PSM/4ASM	16-level PSM, 4-level Amplitude Shift Modulation	750

6. IDENTIFICATION

Call signs are exchanged during linking, and each station identifies every few minutes, all automatically.

7. CLOVER CONTROL BLOCK (CCB)

Each transmission uses a CLOVER Control Block (CCB) to provide synchronization and mode information. Only one CCB is sent per transmitter-ON cycle. The CCB is always sent using a 17-octet block size and a 60% Reed-Solomon encoder efficiency. The CCB uses 2DPSM modulation in the FEC mode and BPSM modulation in the ARQ mode. The CCB is followed by one or more Error Corrector Blocks (ECBs) of data. In FEC mode, the CCB serves as a preamble to the data block which announces the modulation format and the sending station call sign. In ARQ mode, the CCB follows transmission of one or more data blocks and is used to announce the modulation format to be used by the other station during its next transmission.

8. ERROR CORRECTION BLOCK (ECB)

The data field contains one or more ECBs of error-correction encoded data. One of the variable parameters in CLOVER modulation is the length of the ECB which can be 17, 51, 85, or 255 octets long, always sent at a fixed channel rate of 31.25 bit/s. This is analogous to packet length. However, in this case, block length and the number of Reed-Solomon correctable errors are proportional, with longer blocks being able to correct more errors without requiring repeat transmissions. The effective data rate varies with the type of modulation and the encoder efficiency, going up to 750 bit/s under optimum conditions.

9. FEC MODE

This mode allows a sending station to transmit data to one or more receiving stations. FEC mode is a one-way transmission that cannot repeat transmissions for error correction or use adaptive waveform selection. Therefore, the sending station must choose a transmitting modulation format in advance and assume that conditions between the sending station and all other stations are adequate for the chosen mode. The Reed-Solomon algorithm is used to provide receive error correction in FEC mode with a 60% code rate. Both BPSM and QPSM modulations use 85-character blocks, and the default modulation format is 2DPSM in FEC mode.

The first CCB frame begins with 2.048 seconds of carrier. A 32-ms carrier-off gap immediately follows this frame and each of the following frames. This first CCB frame becomes identical to all of the following (recurrent) CCB frames from this point on. The next frame sent is a 64-bit synchronization sequence which also lasts 2.048 seconds. All subsequent blocks are immediately preceded by a 32-ms reference tone pulse sequence. The CCB is next, and it is followed by from 3 to 9 ECBs.

10. ARQ MODE

ARQ is a two-way point-to-point mode which provides fully adaptive and error-corrected communications between two stations that are linked together. As in the case of FEC, a varying number of ECBs are sent in each ARQ time frame. The number of ECBs and other timing parameters are adjusted so that the total time for each ARQ frame is exactly 19.488 seconds, regardless of modulation waveform combination used. The full advantages of adaptive waveform control and error correction via repeat transmission are provided to these two stations. Data is communicated between both ARQ stations by adding a series of ECBs of data following the CCB. Although the CCB's waveform parameters remain fixed, the waveform of the ECBs is adaptively adjusted to match current propagation conditions. The throughput rate during ECB transinissions is generally much higher than that used for the CCB, because the ECB uses longer blocks and high-rate modulation waveforms to expedite data transfer. All ARQ link maintenance operations are performed at the CCB level.

ARQ ECBs are always 255 bytes long. The Reed-Solomon code rate is set for 150, 188, or 226 8-bit data bytes per ECB depending upon the ARQ bias selected (Robust, Normal, or Fast, respectively). The default bias setting is Robust. Both stations send CCB frames, which last 2.720 seconds. A limited amount of data may be exchanged within the CCBs (called Chat Mode), although large quantities of data are transferred through the ECBs. Unlike packet radio, CLOVER selectively repeats only those blocks which fail Reed-Solomon correction, not all blocks following a failed block. ARQ is an adaptive mode that does not use 2DPSM modulation.

11. MONITORING CAPABILITY

LISTEN mode permits additional stations to monitor all traffic between linked ARQ mode stations. Call signs of all CLOVER stations being monitored are also displayed, since CLOVER stations automatically identify every few minutes.

CLOVER REFERENCES

Ford, Steve: "HAL Communications PCI-4000 CLOVER-II Data Controller" (Product Review), *QST*, American Radio Relay League, Newington, CT, May 1993, pp. 71-73.

HAL Communications: "CLOVER Glossary," Engineering Document E2000, Rev. D, HAL Communications Corp., Urbana, IL, November 1992.

HAL Communications: "PCI-4000/CLOVER-II Interface Specifications," Engineering Document E2001, Rev. G, HAL Communications Corp., Urbana, IL, March 1993.

HAL Communications: "PCI-4000 CLOVER-II Data Modem Reference Manual" and "PC-CLOVER Operator's Manual," HAL Communications Corp., Urbana, IL, November 1992.

Henry, George W. and Ray C. Petit: "CLOVER - Fast Data on HF Radio," *CQ*, CQ Communications, Hicksville, NY, May 1992, pp. 40-44.

Henry, George W. and Ray C. Petit: "HF Radio Data Communication: CW to CLOVER," *Communications Quarterly*, CQ Communications, Hicksville, NY, Spring 1992, pp. 11 -24.

Horzepa, Stan (ed); George W. Henry and Ray C. Petit: "CLOVER Development Continues," "Gateway," *QEX*, American Radio Relay League, Newington, CT, March 1992, pp. 12-14.

Petit, Ray C.: "CLOVER is Here," *RTTY Journal*, Fountain Valley, CA; January 1991, pp. 16-18; February 1991, pp. 12-13; March 1991, pp. 16-17; April 1991, p.10.

Petit, Ray C.: "CLOVER Status Report," *RTTY Journal*, Fountain Valley, CA, January 1992, pp. 8-9.

Petit, Ray C.: "The 'CLOVERLEAF' Performance-Oriented HF Data Communication System," *QEX*, American Radio Relay League, Newington, CT, July 1990, pp. 9-12.

Townsend, Jay: "CLOVER - PCI-4000," *RTTY Journal*, Fallbrook, CA, April 1993, pp. 3-4; May/June 1993, p. 20.

CLOVER-2000

1. INTRODUCTION & WAVEFORM

CLOVER-2000 is a faster version of CLOVER (about four times faster) that uses eight tone pulses, each of which is 250 Hz wide, spaced at 250-Hz centers, contained within a 2 kHz bandwidth between 500 and 2,500 Hz. The eight tone pulses are sequential, with only one tone being present at any instant and each tone lasting 2 ms. Each frame consists of eight tone pulses lasting a total of 16 ms, so the base modulation rate of a CLOVER-2000 signal is always 62.5 symbols per second (regardless of the type of modulation being used). Its CCIR emission designation is 2K0H J2 DEN or 2K0H J2 BEN and it has a 2:1 voltage (6 dB power) crest factor. CLOVER-2000's maximum raw data rate is 3,000 bits per second. Allowing for overhead, CLOVER-2000 can deliver error-corrected data over a standard HF SSB radio channel at up to 1,994 bits per second, or 249 characters (8-bit bytes) per second. These are the uncompressed data rates; the maximum throughput is typically doubled for plain text if compression is used.

2. CLOVER SIMILARITIES

CLOVER-2000 is similar to the previous version of CLOVER. The transmission protocols and Reed-Solomon error detection and correction algorithm sections in the original CLOVER technical description are therefore not repeated here. The original descriptions of the CLOVER Control Block (CCB) and Error Correction Block (ECB) still apply for CLOVER-2000, except for the higher data rates inherent to CLOVER-2000. Just like CLOVER, all data sent via CLOVER-2000 is encoded as 8-bit data bytes and the error-correction coding and modulation formatting processes are transparent to the data stream — every bit of source data is delivered to the receiving terminal without modification. Control characters and special "escape sequences" are not required or used by CLOVER-2000. Compressed or encrypted data may therefore be sent without the need to insert (and filter) additional control characters and without concern for data integrity.

3. SIGNAL CHARACTERISTICS

Five different types of modulation may be used in the ARQ mode - BPSM (Binary Phase Shift Modulation), QPSM (Quadrature PSM), 8PSM (8-level PSM), 8P2A (8PSM + 2-level Amplitude Shift Modulation), and 16P4A (16 PSM plus 4 ASM). The same five types of modulation used in ARQ mode are also available in Broadcast (FEC) mode, with the addition of 2-Channel Diversity BPSM (2DPSM). Each CCB is sent using 2DPSM modulation, 17 byte block size, and 60% bias.

DATA THROUGHPUT
in bits per second (bps)

	BIAS (Error Corrector Efficiency)			
	Robust		*Normal*	*Fast*
	(60%)		(75%)	(90%)
	FEC	ARQ	ARQ	ARQ
	Mode	Mode	Mode	Mode
Modulation	(bps)	(bps)	(bps)	(bps)
16P4A	1334	1323.5	1658.8	1994.1
8P2A	905	882.4	1105.9	1329.4
8PSM	741	661.8	829.4	997.1
QPSM	427	441.2	552.9	664.7
BPSM	224	220.6	276.5	332.4
(FEC only) 2DPSM	108	—	—	—

The maximum ARQ data throughput varies from 336 bits per second for BPSM to 1992 bits per second for 1 6P4A modulation. BPSM is most useful for weak and badly distorted data signals while the highest format (16P4A) needs extremely good channels, with high SNRs and almost no multipath.

4. HARDWARE IMPLEMENTATONS

Two different CLOVER-2000 modems are available from HAL Communications, the PCI-4000/2K and the DSP-4100/2K. The PCI-4000/2K is for use inside dedicated desk-top personal computers. The PCI-4000/2K may be installed in any IBM-compatible personal computer that uses an 80386 or faster microprocessor (386, 486, Pentium, etc.) and supports the ISA PC plug-in card bus. The DSP-4100/2K is for connection to lap-top or non-IBM-PC data systems. The DSP-4100/2K is a standalone DSP modem that may be used with any computer or data terminal equipment having an RS-232 port.

5. BI-DIRECTIONAL ARQ

Most ARQ protocols designed for use with HF radio systems can send data in only one direction at a time. For example, when using CCIR-476/625 (SITOR) or PACTOR, one station sends all of its data, ending the transmission with an "OVER" command. The second station may then send its information. Because CLOVER-2000 does not need an "OVER" command, data may flow in either direction at any time. The CLOVER ARQ time frame automatically adjusts to match the data volume to be sent in either or both directions. When first linked, both sides of the ARQ link exchange information using six bytes of the CCB. When one station has a large volume of data buffered and ready to send, ARQ mode automatically shifts to an expanded time frame during which one or more 255 byte data blocks are sent. If the second station also has a large volume of data buffered and ready to send, its half of the ARQ frame is also expanded. Either or both stations will shift back to CCB level when all buffered data has been sent. This feature provides the benefit of full-duplex data transfer but requires use of only simplex frequencies and half-duplex radio equipment. This two-way feature of CLOVER can also provide a back-channel order-wire capability. Communications may be maintained in this "chat" mode at 55 words per minute, which is more than adequate for real-time keyboard-to-keyboard communications.

6. BINARY FILE TRANSFER USING CLOVER-2000

The effective data throughput rate of CLOVER-2000 can be even higher when binary file transfer mode is used with data compression. The binary file transfer protocol used by HAL Communications operates with a terminal program explained in the HAL E2004 Engineering Document listed under References. Data compression algorithms tend to be context sensitive-compression that works well for one mode (e.g. text), may not work well for other data forms (graphics, etc.). The HAL terminal program uses the PK-WARE compression algorithm which has been proven to be a good general-purpose compressor for most computer files and programs. Other algorithms may be much more efficient for some data formats, particularly for compression of graphic image files and digitized voice data. The HAL Communications PCI-4000/2K and DSP-4100/2K modems can be operated with other data compression algorithms in the users' computers.

CLOVER 2000 REFERENCES

HAL Communications: "CLOVER-2000 High Performance Protocol," **http://www.halcomm.com/clvr2k**.
HAL Communications: "Binary File Transfer Protocol," E2004, Rev. B, HAL Communications Corp., Urbana, IL, December 1997.
HAL Communications: "CLOVER-2000 Waveform & Protocol," E2007, Rev. C, HAL Communications Corp., Urbana, IL, May 1999.
Wickwire, Ken, Mike Bernock and Bob Levreault, "On-air Measurements of CLOVER II and CLOVER 2000 Throughput," *Proc. 17th ARRL/TAPR Digital Communications Conference*, ARRL, Newington, CT, 1998, pp. 120-130.

G-TOR

1. PROTOCOL

G-TOR* (Golay-Teleprinting Over Radio) can be viewed, in part, as a variant of the Automatic Link Establishment (ALE) protocol, outlined in MIL-STD-188-141 A. G-TOR combines the error correcting properties of ALE, including Forward Error Correction (FEC) coding and full-frame interleaving, the Automatic Repeat reQuest (ARQ) cycle of Packet and a new application of the invertibility of the Golay code, to produce a faster new mode.

2. DATA FRAME STRUCTURE

G-TOR is a synchronous transmission system with a data frame duration of 1.92 seconds and a 0.48-second window between data fraines, for a total cycle time of 2.40 seconds regardless of transmission rate. Data frames are 192, 384, or 576 bits long sent at 100, 200, or 300 symbols/sec, respectively, with the data rate dependent on band conditions. Each data frame consists of a Data field and Status byte, followed by a two-byte Cyclic Redundancy Check (CRC). No start or ending flags are added to any of the frames, thus lowering overhead and resulting in improved frame efficiency relative to AMTOR and PacTOR. The Data field contains 21, 45, or 69 eight-bit bytes sent at 100, 200, or 300 symbols/sec, respectively. The Status byte provides the frame number identification, data format (whether standard 8-bit ASCII or Huffman compressed), and a command (data, turnaround request, disconnect, or connect) for a total of 8 bits.

3. ACKNOWLEDGEMENT (ACK) FRAME STRUCTURE

ACK frames are used to acknowledge correct or incorrect receipt of data frames, to request changes in transmission speed, and to change the direction of information flow. There are five different ACK frames: Data frame received without error (send next frame), Data frame error detected, Speed-up, Speed-down, and Changeover. Each of the ACK frames consists of two eight-bit bytes sent from the information-receiving station to the information- sending station at 100 symbols/sec, for a duration of 0.16-second during the 0.48-second window between data frames. The Changeover ACK frame initiates a changeover in information flow

*G-TOR is a trademark of Kantronics, Inc.

direction by starting out with a two-byte Changeover ACK (which causes the information-sending station to stop sending) followed by 19 data bytes, a single status byte, and a two-byte CRC, for a duration of 1.92 seconds (the same as a data frame). None of the ACK frames are interleaved; however, each is generated from a set of pseudorandom numbers and up to three bit-errors are allowed per ACK, thus reducing needless retransmissions from faulty ACK signals. Hence the ACKs are called fuzzy. Link quality, denoted by a set number of consecutive good or bad frames, determines link speed.

4. ASCII CHARACTERS AND HUFFMAN / RUN-LENGTH COMPRESSION

G-TOR frames are sent in normal ASCII or are Huffman and run-length encoded, depending upon which is more efficient on a frame-by-frame basis. The Huffman table for G-TOR is unique: It differs from the PacTOR table in that it emphasizes English over German character usage and upper and lower case characters are swapped automatically (frame-by-frame) in a third attempt to compress data—hence Huffman forms A and B.

5. GOLAY ERROR-CORRECTION CODING AND INTERLEAVING

G-TOR uses extended Golay coding which is capable of correcting three or fewer errors in a received 24-bit code word. The Golay code used in G-TOR is a half-rate code, so that the encoder generates one error-correction bit (a parity bit) for every data transmitted. Interleaving is also used to correct burst errors which often occur from lightning, other noise, or interference. Interleaving is the last operation performed on the frame before transmission and de-interleaving is the first operation performed upon reception. Interleaving rearranges the bits in the frame so that long error bursts can be randomized when the de-interleaving is performed. When operating at 300 symbols/second, the interleaver reads 12-bit words into registers by columns and reads 48-bit words out of the registers by rows. The de-interleaver performs the inverse, reading the received data bits into registers by rows and extracting the original data sequence by reading the columns. A long burst of errors, for example 12-bits in duration, will be distributed into 48 separate 12-bit words before the error correction process is applied. This effectively nullifies the errors. Both data frames and parity frames are completely interleaved. In addition, by using the invertibility characteristic of Golay code words, data frames are always alternated with data frames coded in Golay parity bits. In this way, G-TOR can maintain full speed (when band conditions are good)—rather than fall to rate-1/2. Receiving parity bits can be used as data or as parity.

6. LINK INITIALIZATION

To establish a link, the information-sending station transmits the call sign of the intended receiver. Once the information-receiving station has synchronized, it sends an ACK to the information-sending station and data transmission begins.

7. SIGNAL CHARACTERISTICS

G-TOR uses frequency-shift keying like PacTOR and packet radio. At 300 symbols/second, and with the recommended frequency shift of 170 or 200 Hz, G-TOR's spectral characteristics are almost identical to those of packet radio.

8. ERROR DETECTION AND ARQ CYCLE

G-TOR provides error correction by using a combination of both ARQ retransmission and forward error-correction. The error-detection code transmitted with each frame is a 2-byte CRC code, the same used in the AX.25 packet protocol, and it is used to determine if the frame was received correctly before error correction is initiated and after error correction is completed, to ensure that the error-correction process has successfully removed all errors in the packet. Although the CRC error-detection code is used on every frame to detect errors, the Golay error-correction procedure is skipped unless errors are detected. This ability to skip unnecessary error correction is extremely valuable since forward error correction is very costly in terms of throughput. The Golay code used in G-TOR is a half-rate code, with one error-correction bit required for every information bit; however, by using the invertibility of the extended Golay code, the half-rate transmission result normally encountered with FEC systems is avoided. Frames made up of parity bits can be fully converted to data frames. Received frames are synchronized, deinterleaved, decoded and checked for proper CRC. If the frame is found to be in error, the information-receiving station will request that the matching parity frame be sent. If the parity (or data) frame that follows is found to be correct, that frame is acknowledged. If, however, it too is in error, it is combined with the previous data (or parity) frame in an attempt to recover the original data bits. In this way the system has three chances to recover the original data from the transmission of one data and one parity frame. If unsuccessful, the ARQ cycle begins again. The dispersal of noise-burst errors via interleaving, combined with the power of the Golay code to correct 3 bits in every 24, usually results in the recovery of error-free frames.

G-TOR REFERENCES

Anderson, Phil: "G-TOR's Evolutionary Improvements!" Kantronics Inc., Lawrence, KS, 1994.

Anderson, Phil: "H F ARQ Protocols," *RTTY Digital Journal*, April 1994.

Anderson, Phil, and Glenn Prescott: "Error-Correcting Codes," *RTTY Digital Journal*, March 1994.

Anderson, Phil, Michael Huslig, Glenn Prescott, and Karl Medcalf: "G-TOR: The New, Faster HIF Digital Mode for the KAM Plus," *RTTY Digital Journal*, March 1994, pp.1-2.

Ford, Steve: "G-TOR vs. PacTOR vs. AMTOR: A Nonscientific Throughput Test," *QEX*, American Radio Relay League, Newington, CT, May 1994, p. 18.

Kantronics: "G-TOR: The New HF Digital Mode for the KAM Plus and KAM Enhancement Board," Kantronics Inc., Lawrence, KS, 1994.

Karn, Phil: "Toward New Link-Layer Protocols," *QEX*, American Radio Relay League, Newington, CT, June 1994, pp. 3-10.

Prescott, Glenn, Phil Anderson, Mike Huslig, and Karl Medcalf: "G-TOR: A Hybrid ARQ Protocol for Narrow Bandwidth HF Data Communication," *QEX*, American Radio Relay League, Newington, CT, May 1994, pp. 12-19.

PACTOR

1. PROTOCOL

PACTOR can be viewed as a combination of two earlier digital modes, packet radio and AMateur Teleprinting Over Radio (AMTOR). PACTOR provides improved throughput because its transmission speed adapts to the quality of the link and it uses Huffman compressed characters. PACTOR operates over half-duplex links and uses an Automatic Repeat reQuest (ARQ) protocol, acknowledging each individual data packet with a short Control Signal (CS). Some PACTOR implementations provide a Memory-ARQ feature to determine and store the relative strength of each received bit. Copies of corrupted frames stored this way are correlated with frames received later, to provide a coding gain for improved error correction.

2. PACKET LENGTH AND ACKNOWLEDGMENT

PACTOR is a synchronous transmission system with a packet duration of 0.96 second, a CS duration of 0.12 second, and an idle time of 0.17 second for a total cycle time of 1.25 seconds. The idle time is required for turaround procedures and settling, which allows for a maximum distance of about 20,000 km as in AMTOR. Clock reference is crystal controlled with an internal standard to at least 15×10^{-6} accuracy. The initiating station is the Master, and the other station is the Slave. PACTOR subjects each received packet to a Cyclic Redundancy Check (CRC) which triggers an ARQ for packets failing the CRC. The receiving station provides acknowledgment by sending a CS. Repetition of the same CS indicates a request for a packet to be repeated. The CS has a standard length of 12 bits and is always sent at 100 bauds.

3. PACKET COMPOSITION AND BAUD RATE

Packets are either 96 bits long sent at 100 bauds or 192 bits long sent at 200 bauds, with the data rate dependent on conditions. Each packet consists of a Header byte, Data field, and Status byte, followed by the CRC byte given twice. The Header byte consists of the 8-bit pattern for 55 hexadecimal and is used for synchronization, MemoryARQ, and listen mode. The Data field contains 64 bits if sent at 100 bauds or 160 bits if sent at 200 bauds. Its format is normally Huffman-compressed ASCII, with conventional 8-bit ASCII as the alternative. The Status byte provides the packet count, data format (whether standard 8-bit ASCII or Huffman-compressed ASCII), break-in request, and QRT bit, for a total of 8 bits. The CRC calculation is based on the ITU-T polynomial $xE16+xE12+xE5+1$. The CRC byte is calculated for the whole packet starting with the data field, without Header, and consists of 16 bits. There are four different Control Signals, identified as CS1, CS2, CS3, and CS4, the functions of which are explained below.

4. LINK INITIALIZATION

To initiate a PACTOR connection, the Master station sends a special synchronization packet which contains the standard Header byte followed by the call sign (address) of the Slave station in both a 100-baud and a 200-baud bit pattern. (Each address field allows for up to eight characters, with the character 0F hexadecimal following the call sign in each unused space.) This allows the 200-baud bit pattern to determine the quality of the channel: The Slave responds with CS1 (4D5 hexadecimal) if the 200-baud bit pattern was received without error, or CS4 (D2C hexadecimal) if not, which leads to an instant reduction of data rate to 100 bands. After receiving the first CS1 or CS4 from the Slave, the Master sends the first data packet with Header=AA hexadecimal and packet count=1. System specific data, including the Master call sign, is sent automatically at the beginning. If the Slave is busy, it can both acknowledge and reject a connection by sending one CS2 (AB2 hexadecimal) each time it receives a correct synchronization pattern. The Master terminates its connect request after receiving CS2 twice in succession.

5. CHANGING SPEED

The decision regarding a speed change is made at the receiving end, by automatically evaluating the data input rate and packet statistics including error rate and number of retries. The receiving station transmits CS1 (acknowledge) following each correctly received packet, or CS4 (speed change request) following receipt of a bad 200-baud packet. The data contained in each unacknowledged 200-baud packet is automatically repeated at 100 bauds. This repetition requires several 100-baud packets because of the smaller Data field. If the receiving station acknowledges a correctly received 100-baud packet with CS4, the transmitting station sends the next packet at 200 bauds. If the following 200-baud packet is not acknowledged after an operator-selectable number of attempts (normally two), the speed is automatically set back to 100 bauds.

6. CHANGING DIRECTION AND ENDING CONNECTION

The receiving station can change to transmit by sending CS3 (3413 hexadecimal) as a break-in request at the head of its first data packet. At the end of a connection, special end of contact (QRT) synchronization packets are transmitted which contain the receiver address in the reversed order. This process is repeated until the sending station has received the acknowledgement.

7. ASCII CHARACTERS AND HUFFMAN COMPRESSION

PACTOR normally uses ASCII characters that have been compressed with a Huffman algorithm. This Huffman compression reduces the average character length for improved efficiency. The Annex shows the Huffman compressed equivalent of each ASCII character used in PACTOR, with the least significant bit (LSB) given first. The length of individual characters varies from 2 to 15 bits, with the most frequently used characters being the shortest. This results in an average character length of 4 to 5 bits for English text, instead of the 8 bits required for normal ASCII.

8. SIGNAL CHARACTERISTICS

PACTOR uses frequency-shift keying (FSK). With the recommended frequency shift of 200 Hz, PACTOR can be received through a filter with a bandwidth as narrow as 600 Hz.

PACTOR REFERENCES

Helfert, Hans-Peter, and Ulrich Strate: "PacTOR Radioteletype with Memory ARQ and Data Compression," *QEX*, American Radio Relay League, Newington, CT, October 1991, pp. 3-6.

Horzepa, Stan: "PacTOR: Better HF Data Communications for the Rest of Us?", *QST*, American Radio Relay League, Newington, CT, February 1993, p. 98.

Rogers, Buck: "PacTOR - The New Frontier," *CQ*, CQ Communications, Hicksville, NY, July 1993, pp. 88-95.

Van Der Westhuizen, Mike: "A Practical Comparison Between Clover and Pactor Data Transfer Rates," *CQ*, CQ Communications, Hicksville, NY, February 1994, pp. 40-42.

PACTOR Huffman Compression

Char	ASCII	Huffman (LSB [sent first] on left)
space	32	10
e	101	011
n	10	0101
i	105	1101
r	114	1110
t	116	00000
s	115	00100
d	100	00111
a	97	01000
u	117	11111
l	108	000010
h	104	000100
g	103	000111
m	109	001011
<CR>	13	001100
<LF>	10	001101
o	111	010010
c	99	010011
b	98	0000110
f	102	0000111
w	119	0001100
D	68	0001101
k	107	0010101
z	122	1100010
.	46	1100100
,	44	1100101
S	83	1111011
A	65	00101001
E	69	11000000
P	112	11000010
v	118	11000011
O	48	11000111
F	70	11001100
B	66	11001111
C	67	11110001
I	73	11110010
T	84	11110100
O	79	000101000
P	80	000101100
1	49	001010000
R	82	110000010
(	40	110011011
)	41	110011100
L	76	110011101
N	78	111100000
Z	90	111100110
M	77	111101010
9	57	0001010010

Char	ASCII	Huffman (LSB [sent first] on left)
W	87	0001010100
5	53	0001010101
y	121	0001010110
2	50	0001011010
3	51	0001011011
4	52	0001011100
6	54	0001011101
7	55	0001011110
8	56	0001011111
H	72	0010100010
J	74	1100000110
U	85	1100000111
V	86	1100011000
<FS>	28	1100011001
x	120	1100011010
K	75	1100110100
?	63	1100110101
=	61	1111000010
q	113	1111010110
Q	81	1111010111
j	106	00010100110
G	71	00010100111
-	45	00010101111
:	58	00101000111
!	33	11110011101
/	47	11110011110
*	42	001010001100
"	34	110001101100
%	37	110001101101
'	39	110001101110
_	95	111100001100
&	38	111100111001
+	43	111100111110
>	62	111100111111
@	64	0001010111000
$	36	0001010111001
<	60	0001010111010
X	88	0001010111011
#	35	0010100011011
Y	89	00101000110101
;	59	11110000110100
\	92	11110000110101
[	91	001010001101000
]	93	001010001101000
<DEL>	127	110001101111000
~	126	110001101111001
}	125	110001101111010
I	124	110001101111011
{	123	110001101111100
`	96	110001101111101
^	94	110001101111110

Char	ASCII	Huffman (LSB [sent first] on left)
<US>	31	110001101111111
<GS>	29	111100001101100
<ESC>	27	111100001101101
<EM>	25	111100001101110
<CAN>	24	111100001101111
<ETB>	23	111100001110000
<SYN>	22	111100001110001
<NAK>	21	111100001110010
<DC4>	20	111100001110011
<DC3>	19	111100001110100
<DC2>	18	111100001110101
<DC1>	17	111100001110110
<DLE>	16	111100001110111
<RS>	30	111100001111000
<SI>	15	111100001111001
<SO>	14	111100001111010
<FF>	12	111100001111011
<VT>	11	111100001111100
<HT>	9	111100001111101
<BS>	8	111100001111110
<BEL>	7	111100001111111
<ACK>	6	111100111000000
<ENQ>	5	111100111000001
<EOT>	4	111100111000010
<ETX>	3	111100111000011
<STX>	2	111100111000100
<SOH>	1	111100111000101
<NUL>	0	111100111000110
<SUB>	26	111100111000111

PACTOR II

1. INTRODUCTION

PACTOR II is a fully backward-compatible improvement to the original PACTOR system. It is an adaptive mode that applies different modulation and encoding methods depending on the channel quality. PACTOR II performs its initial link setup using PACTOR in order to achieve compatibility. An automatic handover to PACTOR II is made only if both stations are capable of PACTOR operation. While PACTOR uses frequency-shift keying (FSK) modulation, PACTOR II uses two-tone differential phase-shift keying (DPSK) modulation.

PACTOR II's highest data transfer rate without compression is 700 bit/s under optimum conditions, which can provide a maximum effective throughput of up to 1200 bit/s (for text with real-time data compression enabled). PACTOR II's improved error control coding enables it to operate on circuits with (S+N)/N ratios as much as 7 dB lower than PACTOR can tolerate.

2. STRUCTURE AND TIMING OF PACTOR II FRAMES

Similar to PACTOR, PACTOR II is also a half-duplex protocol that uses Automatic Repeat reQuest (ARQ) to acknowledge each individual data packet with a short control signal (CS). An "over" must be inserted into the data stream to change direction. The basic PACTOR 11 frame structure consists of a header, a variable data field, a status byte and a cyclic redundancy check (CRC). The standard cycle duration for PACTOR II is the same 1.25 s used by PACTOR, with the same 0.17s idle time (for propagation delay), to maintain interoperability. Since PACTOR II uses a longer CS of 0.28 s (instead of PACTOR's 0.12 s), the packet length had to be shortened to 0.8 s (from PACTOR's 0.96 s). The transceiver requirements for transmit delay and receiver recovery time therefore remain the same for PACTOR II as they are for PACTOR.

PACTOR II uses six different CSs, each consisting of 40 bits. As in PACTOR, CS1 and CS2 are used to acknowledge/request packets and CS3 forces a break-in. CS4 and CS5 handle the speed changes, and CS6 is a toggle for the packet length.

Due to the signal propagation delay and equipment switching delays PACTOR II has a maximum range for ARQ contacts of approximately 20,000 km. As with PACTOR, a long-path option is available to enable contacts up to 40,000 km. The sending station must call the receiving station in "long-path" mode. Initial contact is established using the ordinary PACTOR protocol, but with a cycle time of 1.4 s instead of 1.25 s. This longer cycle time allows for the much greater propagation delays found on long-path contacts. The link then automatically switches to PACTOR II, with the same cycle duration. A switch between "normal" and "long-path" modes requires ending the link and reconnecting. In the new "data mode" (see below), timing is also automatically adjusted to obtain longer receiving gaps.

Unlike PACTOR, PACTOR II automatically switches to longer packets if the data blocks are not filled up with idles; i.e., if the transmitter buffer indicates that more information has to be transferred than fits into the standard packets. If the information sending station (ISS) prefers to use long packets, it sets the long-cycle flag in the status word. The PACTOR II information receiving station (IRS) then finally can accept the proposed change of the cycle duration by sending a CS6. This situation, for example, occurs when reading longer files out of mailboxes. The long packets are basically made up like the short ones, but consist of a larger data field that may contain up to 2208 bits of usable information. The length of these data packets is 3.28 s, which leads to an entire cycle duration of 3.75 s in this "data mode."

3. ERROR CONTROL CODING

Effective transmission of data over difficult HF paths requires substantial error control coding (ECC). Full frame interleaving is required for cancelling out error bursts and short fading periods, especially with the longer data packets being used. PACTOR II uses a convolutional code with a constraint length of 9, a real Viterbi decoder, and soft decision. Due to the high coding gain and resulting capacity of error correction without requesting a repetition of the entire packet, a significant increase in the effective throughput can usually be attained.

ECC requires redundant bits be appended to the data before transmission through a noisy channel. The redundant bits are generated from the original data by applying special rules that depend on the chosen code. The ratio of the number of information bits to the whole length is the code rate.

Two main approaches of ECC can be distinguished: block codes and convolutional codes. Convolutional codes require data interleaving to be effective on channels with burst errors from noise or interference. When applying block codes (such as Golay, Hamming and Reed-Solomon), the message or packet is divided into data blocks. Each block is then encoded separately and forms a code word. Block codes can easily be implemented as they often show a cyclic property and may not require much processing power. Block codes do not require interleaving because of their tolerance for burst errors, thus are efficient for use over HF radio.

Convolutional codes encode the entire message packet and the resulting code words are longer than the original packet. The complexity of a convolutional code mainly depends on the number of tapped shift registers, which work as binary convolvers, and represent the heart of the convolutional encoder. The eight shift registers used in PACTOR II provide a constraint length of 9. This sets the upper boundary for the achievable coding gain. PACTOR II uses a Viterbi decoder with soft decision to implement maximum likelihood decoding.

PACTOR II uses a rate 1/2 convolutional code. Code with higher rates, e.g., rate 2/3 and rate 7/8, are derived by a process called "puncturing." Prior to their transmission, certain of the symbols of the rate 1/2 encoded stream are punctured or deleted and not transmitted. At the receiving end, the punctured encoded bits are replaced with "null" symbols prior to decoding with the rate 1/2 decoder. The decoder treats these null symbols neither as a received "1" nor "0" but as an exactly intermediate value. No information is thus conveyed by that symbol that may influence the decoding process. The major advantage of this approach is that a single code rate decoder (in this case a rate 1/2 decoder) can implement a wide range of codes. In PACTOR II, the Viterbi algorithm is implemented for decoding the convolutional code.

PACTOR II provides a more robust Listen Mode than PACTOR because just the short header has to be received correctly to enable use of the ECC. Burst errors may be corrected by monitoring stations. The same speed and encoding levels used by PACTOR II are also available in its Unproto Mode. On the receiving side, the correct mode is detected automatically.

4. MODULATION

PACTOR II uses various DPSK modulation schemes and different code rates, depending on conditions. Differential binary phase-shift keying (DBPSK) is the most robust modulation used by PACTOR II, so it is used for sending all control signals. If propagation conditions improve, PACTOR II automatically switches to differential quadrature phase-shift keying (DQPSK) modulation. If conditions improve further, it changes to 8-DPSK and finally to 16DPSK under optimum conditions.

A single-tone 100 bit/s DBPSK signal can transfer 100 bit/s. PACTOR II uses two-tone DBPSK modulation in its most robust mode, which can be thought of as two single-tone DBPSK signals used together, resulting in a data rate 200 bit/s. This data rate is lowered by the redundancy required in the coding to reduce the error rate, so that the effective throughput becomes 100 bit/s with rate 1/2 coding.

PACTOR II uses different forms of two-tone DPSK modulation in all of its modes and starts using DBPSK modulation at a data rate of 200 bit/s with rate 1/2 coding to provide a throughput of 100 bit/s. The next step uses DQPSK modulation at a data rate of 400 bit/s also with rate 1/2 coding, for a throughput of 200 bit/s. This is followed by 8-DPSK modulation at a data rate of 600 bit/s with rate 2/3 coding, for a throughput of 400 bit/s. Finally, in the best propagation conditions, PACTOR II uses 16-DPSK modulation at 800 bit/s with a rate 7/8 coding, for a throughput of 700 bit/s. These throughputs are all the maximum uncompressed data transfer rates which are realised only if no packets need to be re-sent because of errors. PACTOR II automatically selects the modulation and hence the data rate by considering the error repeat history (although this is not a true link quality assessment). When data compression is active, the effective data transfer rates may be considerably higher.

5. ON-LINE DATA COMPRESSION

Both PACTOR and PACTOR II use an automatic online data compression algorithm that is optimised for text, starting with Huffman encoding. Additionally, PACTOR II uses run-length encoding and pseudo-Markov compression (PMC). Compared to 8-bit ASCII (plain text), PMC yields a compression factor of around 1.9. This leads to an effective speed of about 600 bit/s in average propagation conditions in data mode. PACTOR II is already around three times faster than PACTOR on average channels. However, the maximum effective speed in good conditions can reach 1200 bit/s. PACTOR II firmware automatically checks whether PMC, Huffman or the original ASCII code is the best choice. PACTOR II is also able to transfer and 7-bit ASCII data and other binary information (such as programs, pictures and voice files) which can be split up into 4-bit nibbles by automatically switching off its on-line data compression.

6. SIGNAL CHARACTERISTICS

Similar to PACTOR, the tones of the PACTOR II signal are spaced at 200 Hz. Their centre frequency may be defined in steps of 1 Hz, by software command, between 400 Hz and 2600 Hz, as long as the shift remains 200 Hz. In the PACTOR II system, the transferred information is swapped from one channel (tone) to the other in every cycle. This swapping provides resistance to strong narrowband interference (e.g., CW) which might completely overpower one channel, so the signal can still get transferred but at a reduced speed. PACTOR II signals can be received through filters as narrow as 500 Hz.

PSK31

1. INTRODUCTION

PSK31 is a digital communications mode which is intended for live keyboard-to-keyboard conversations, similar to radioteletype. Its data rate is 31.25 baud (about 50 word-per-minute), and its narrow bandwidth (less than 80 Hz) reduces its susceptibility to noise. It uses BPSK modulation without error correction or QPSK modulation with error correction (convolutional encoding and Viterbi decoding). In order to eliminate splatter from the phase-reversals inherent to PSK, the output is cosine-filtered before reaching the transmitter audio input. PSK31 is readily monitored and the most popular implementation uses DSP software running on a computer soundcard inside an IBM PC-compatible computer.

There is a preamble at the start of each transmission and a postamble at the end. The preamble is an idle signal of continuous zeroes, corresponding to continuous phase reversals at the symbol rate of 31.25 reversals/second. The postamble is just continuous unmodulated carrier, representing a series of logical ones. This makes it possible to use the presence or absence of the reversals to "squelch" the decoder so that the screen doesn't fill with noise when there is no signal.

2. VARICODE CHARACTERS

Different characters are represented by a variable-length combination of bits called Varicode. Because shorter bit-lengths are used for the more common letters, Varicode improves efficiency in terms of the average character duration. Varicode is also self-synchronizing: No separate process is needed to define where one character ends and the next begins, since the pattern used to represent a gap between two characters (at least two consecutive zeroes) never occurs in a character. Because no Varicode characters can begin or end with a zero, the shortest character is a single one by itself. The next is 11, then 101, 111, 1011, and 1101, but not 10, 100, or 1000 (because they end with zeroes), and not 1001 (since it contains two consecutive zeros). This scheme generates the 128-character ASCII set with ten bits.

The Varicode character set is shown following, starting with NUL and ending with DEL. The codes are transmitted left bit first, with "0" representing a phase reversal on BPSK and "1" representing a steady carrier. A minimum of two zeros is inserted between characters. Some implementations may not handle all the codes below 32. Note that the lower case letters have the shortest patterns and so are the fastest to transmit.

The Varicode Character Set

NUL	1010101011		DLE	1011110111
SOH	1011011011		DCI	1011110101
STX	1011101101		DC2	1101101101
ETX	1101110111		DC3	1101101111
EOT	1011101011		DC4	1101011011
ENQ	1101011111		NAK	1101101011
ACK	1011101111		SYN	1101101101
BEL	1011111101		ETB	1101010111
BS	1011111111		CAN	1101111011
HT	11101111		EM	1101111101
LF	11101		SUB	1110110111
VT	1101101111		ESC	1101010101
FF	1011011101		FS	1101011101
CR	11111		GS	1110111011
SO	1101110101		RS	1011111011
SI	1110101011		US	1101111111
SP	1		C	10101101
!	111111111		D	10110101
"	101011111		E	1110111
#	111110101		F	11011011
$	111011011		G	11111101
%	1011010101		H	101010101
&	1010111011		I	1111111
'	101111111		J	111111101
(	11111011		K	101111101
)	11110111		L	11010111
*	101101111		M	10111011
+	111011111		N	11011101
,	1110101		O	10101011
-	110101		p	11010101
.	1010111		Q	111011101
/	110101111		R	10101111
0	10110111		S	1101111
1	10111101		T	1101101
2	11101101		U	101010111
3	11111111		V	110110101
4	101110111		X	101011101
5	101011011		Y	101110101
6	101101011		Z	101111011
7	110101101		[	1010101101
8	110101011		\	111110111
9	110110111		]	111101111
:	11110101		^	111111011
;	110111101		_	1010111111
<	111101101		.	101101101
=	1010101		/	1011011111
>	111010111		a	1011
?	1010101111		b	1011111
@	1010111101		c	101111
A	1111101		d	101101
B	11101011		e	11
f	111101		s	10111
g	1011011		t	101
h	101011		u	110111

i	1101	v	1111011
j	111101011	w	1101011
k	10111111	x	11011111
l	11011	y	1011101
m	111011	z	111010101
n	1111	{	1010110111
o	111	|	110111011
p	1111111	}	1010110101
q	110111111	~	1011010111
r	10101	DEL	1110110101

3. QPSK MODE

The QPSK mode reduces the error-rate while keeping the bandwidth and the traffic speed the same. There is a 3-dB SNR penalty with QPSK, because the same transmitter power is being shared by twice the signals. Therefore, the error-correction scheme has to be at least good enough to correct the extra errors which result from the 3 dB SNR penalty, and preferably a lot more, or it will not be worth doing. By doing simulations in a computer, and tests on the bench with a noise generator, it has been found that when the bit error-rate is less than 1% with BPSK, it is much better than 1% with QPSK and error-reduction, but when the BER is worse than 1% on BPSK, the QPSK mode is actually worse than BPSK. Therefore, if we are dealing with radio paths where the signal is just simply very noisy, there is actually no advantage to QPSK at all!

On-the-air testing shows that QPSK with the convolutional coding for error-reduction is usually better than BPSK, except where the signal was deliberately attenuated to make it artificially weak. Typical radio circuits are far from being non-fading with white noise. Typical radio paths have errors in bursts rather than randomly spread, and error-reduction schemes can give useful benefits in this situation in a way that cannot be achieved by anything which can be done in the linear part of the signal path. With the convolutional coding used in PSK31, a 5:1 improvement is typical, but it does depend on the kind of path being used. There may be times when one mode works better than the other, and other times when the reverse will be the case. The switch between "straight" BPSK and "error-corrected" QPSK modes in PSK31 is done with both the bandwidth and the data-rate remaining the same. Contacts tend to start on BPSK and change to QPSK if both stations agree. Although both stations have to be using the same sideband in QPSK, it doesn't matter for BPSK.

4. CONVOLUTIONAL CODING

Convolutional coding is used to reduce errors in the QPSK mode. In a convolutional code, the characters are converted to a bitstream and then this bitstream is itself processed to add the error-reduction qualities. Since the convolutional code used in PSK31 doubles the number of data bits, it is a natural choice for the QPSK mode which provides double the bit-rate available with BPSK. The convolutional encoder generates one of the four phase-shifts, not from each data bit to be sent, but from a sequence of them. This means that each bit is effectively spread out in time, intertwined with earlier and later bits in a precise way. The more spread out, the better will be the ability of the code to correct bursts of noise, but too great a spread would introduce an excessive transmission delay. A time spread of 5 bits was chosen.

It is not quite correct to refer to the convolutional code system as "error-correcting" since the raw data is not actually transmitted in its original form and therefore it makes no sense to talk about it being corrupted by the link and corrected in the decoder. In PSK31, the raw data is transformed from binary (1 of 2) to quaternary (1 of 4) in such a way that there is a precisely known pattern in the sequence of quaternary symbols. In the code used in PSK31, each quaternary symbol transmitted is derived from a run of 5 consecutive data bits. This means that each binary bit to be transmitted generates a 5-symbol sequence, overlapping with the sequences from adjacent bits, in a predictable way which the receiver can use to estimate the correct sequence even in the presence of corruptions in parts of the sequence.

The Convolutional Code

The left columns in the following table contain the 32 combinations of a run of five Varicode bits, transmitted left bit first. The right columns are the corresponding phase shifts to be applied to the carrier, in degrees. A continuous phase advance is the same as an HF frequency shift.

00000	180	01000	0	10000	+90	11000	−90
00001	+90	01001	−90	10001	180	11001	0
00010	−90	01010	+90	10010	0	11010	180
00011	0	01011	180	10011	−90	11011	+90
00100	−90	01100	+90	10100	0	11100	180
00101	0	01101	180	10101	−90	11101	+90
00110	180	01110	0	10110	+90	11110	−90
00111	+90	01111	−90	10111	180	11111	0

For example, consider the "space" symbol—a single 1 preceded and followed by character gaps of five zeroes each: 00000 100000. Overlaying a five-bit-wide window on 00000100000, and sliding it from left to right (one bit at-a-time) is illustrated in the following table.

											Phase
0	0	0	0	0	1	0	0	0	0	0	
0	0	0	0	0							180
	0	0	0	0	1						+90
		0	0	0	1	0					−90
			0	0	1	0	0				−90
				0	1	0	0	0			0
					1	0	0	0	0		+90
						0	0	0	0	0	180

Representing 00000100000 would be the successive run-of-five groups 00000, 00001, 00010, 00100, 01000, 10000, 00000. This results in the transmitter sending the QPSK pattern 180, +90, −90, −90, 0, +90, 180. Note that a continuous sequence of zeros (the idle sequence) gives continuous reversals, the same as BPSK.

5. VITERBI DECODING

Viterbi decoding is used on the receiving side. It consists of a whole bank of parallel encoders, each fed with one possible "guess" at the transmitted data sequence. The outputs of these parallel encoders are all compared with the received symbol stream. Each time a new symbol is received, the encoders need to add an extra bit to their sequence guesses and consider that the new bit might be a 0 or a 1. This doubles the number of sequence guesses, but a clever technique allows half of all the guessed sequences to be discarded as being less likely than the other half, and this means that the number of guesses being tracked stays constant. After a large number of symbols have been received, the chances of a wrong guess at the first symbol tends to zero, so the decoder can be pretty sure that the first bit was right and it can be fed to the output. In practice this means that the decoder always outputs decoded data bits some time after they have been received. The one-way delay in PSK31 is 25 bits (800 mS) which is long enough to make sure that the decoder has done a good job, but not so long that it introduces an unacceptable delay in displaying the received text.

6. BIBLIOGRAPHY

The "official" PSK31 Web site is **http://aintel.bi.ehu.es/psk31.html**, operated by Eduardo Jacob, EA2BAJ. The PSK31 software program may be downloaded from this site.

Steve Ford, WB8IMY, "PSK31—Has RTTY's Replacement Arrived?" *QST*, May 1999, pp 41-41.

Peter Martinez, G3PLX, "PSK31: A New Radio-Teletype Mode, *Radio Communication*, December 1998 and January 1999.

"PSK31 Gets Raves," *The ARRL Letter*, Vol. 18, No. 7, February 12, 1999.

About the ARRL

The seed for Amateur Radio was planted in the 1890s, when Guglielmo Marconi began his experiments in wireless telegraphy. Soon he was joined by dozens, then hundreds, of others who were enthusiastic about sending and receiving messages through the air—some with a commercial interest, but others solely out of a love for this new communications medium. The United States government began licensing Amateur Radio operators in 1912.

By 1914, there were thousands of Amateur Radio operators—hams—in the United States. Hiram Percy Maxim, a leading Hartford, Connecticut, inventor and industrialist saw the need for an organization to band together this fledgling group of radio experimenters. In May 1914 he founded the American Radio Relay League (ARRL) to meet that need.

Today ARRL, with approximately 170,000 members, is the largest organization of radio amateurs in the United States. The ARRL is a not-for-profit organization that:
- promotes interest in Amateur Radio communications and experimentation
- represents US radio amateurs in legislative matters, and
- maintains fraternalism and a high standard of conduct among Amateur Radio operators.

At ARRL headquarters in the Hartford suburb of Newington, the staff helps serve the needs of members. ARRL is also International Secretariat for the International Amateur Radio Union, which is made up of similar societies in 150 countries around the world.

ARRL publishes the monthly journal *QST*, as well as newsletters and many publications covering all aspects of Amateur Radio. Its headquarters station, W1AW, transmits bulletins of interest to radio amateurs and Morse code practice sessions. The ARRL also coordinates an extensive field organization, which includes volunteers who provide technical information for radio amateurs and public-service activities. In addition, ARRL represents US amateurs with the Federal Communications Commission and other government agencies in the US and abroad.

Membership in ARRL means much more than receiving *QST* each month. In addition to the services already described, ARRL offers membership services on a personal level, such as the ARRL Volunteer Examiner Coordinator Program and a QSL bureau.

Full ARRL membership (available only to licensed radio amateurs) gives you a voice in how the affairs of the organization are governed. ARRL policy is set by a Board of Directors (one from each of 15 Divisions). Each year, one-third of the ARRL Board of Directors stands for election by the full members they represent. The day-to-day operation of ARRL HQ is managed by an Executive Vice President and his staff.

No matter what aspect of Amateur Radio attracts you, ARRL membership is relevant and important. There would be no Amateur Radio as we know it today were it not for the ARRL. We would be happy to welcome you as a member! (An Amateur Radio license is not required for Associate Membership.) For more information about ARRL and answers to any questions you may have about Amateur Radio, write or call:

ARRL—The national association for Amateur Radio
225 Main Street
Newington CT 06111-1494
Voice: 860-594-0200
Fax: 860-594-0259
E-mail: **hq@arrl.org**
Internet: **www.arrl.org/**

Prospective new amateurs call (toll-free):
800-32-NEW HAM (800-326-3942)
You can also contact us via e-mail at **newham@arrl.org**
or check out *ARRLWeb* at **www.arrl.org/**

INDEX

FEEDBACK

Please use this form to give us your comments on this book and what you'd like to see in future editions, or e-mail us at **pubsfdbk@arrl.org** (publications feedback). If you use e-mail, please include your name, call, e-mail address and the book title, edition and printing in the body of your message. Also indicate whether or not you are an ARRL member.

Where did you purchase this book?
☐ From ARRL directly   ☐ From an ARRL dealer

Is there a dealer who carries ARRL publications within:
☐ 5 miles   ☐ 15 miles   ☐ 30 miles   of your location?   ☐ Not sure.

License class:
☐ Novice   ☐ Technician   ☐ Technician Plus   ☐ General   ☐ Advanced   ☐ Extra

---

Name _____

Daytime Phone (   ) _____

Address _____

City, State/Province, ZIP/Postal Code _____

If licensed, how long? _____

Other hobbies_____

_____

Occupation _____

ARRL member? ☐ Yes ☐ No

Call Sign _____

Age _____

E-mail_____

For ARRL use only	HF DIG
Edition	2 3 4 5 6 7 8 9 10 11 12
Printing	1 2 3 4 5 6 7 8 9 10 11 12

From _____

_____

_____

EDITOR, HF DIGITAL HANDBOOK
ARRL—THE NATIONAL ASSOCIATION FOR AMATEUR RADIO
225 MAIN STREET
NEWINGTON CT 06111-1494

— — — — — — — — — — — please fold and tape — — — — — — — — — — — —